The Evolution of Flight

Georg Glaeser • Hannes F. Paulus • Werner Nachtigall

The Evolution of Flight

Georg Glaeser
Department of Geometry
University of Applied Arts Vienna
Vienna, Austria

Hannes F. Paulus
Universität Wien Integrative Zoologie
Wien, Austria

Werner Nachtigall
Saarland University
Scheidt, Germany

ISBN 978-3-319-57023-5   ISBN 978-3-319-57024-2   (eBook)
DOI 10.1007/978-3-319-57024-2

Library of Congress Control Number: 2017941197

Printed on acid-free paper

This Springer imprint is published by Springer Nature
The registered company is Springer International Publishing AG
The registered company address is: Gewerbestrasse 11, 6330 Cham, Switzerland

Foreword

A trio of authors

This book is organized similarly to its predecessor, "The Evolution of the Eye". Once again, the following statement applies: Animal evolution itself cannot be captured in photographs, but its results certainly can. In this book, we have focused on the evolution of flying, various forms of which have independently been "invented" in the animal kingdom. Mathematician and passionate animal photographer Georg Glaeser has joined forces with the experienced evolutionary biologist Hannes Paulus and the exercise physiologist and flight biophysicist Werner Nachtigall in order to approach this topic with words and pictures in a way that is both generally comprehensible and scientifically sound.

Naturally, the book will not only consist of purely biological accounts, but will also touch upon aspects of Technical Biology. The discipline of Technical Biology describes and explains natural forms and processes in light of physical-technical expertise. Especially in gaining a better understanding of animal flight, certain viewpoints and parameters derived from the physics of technical flight can prove particularly helpful, even though they will only be applied in this book to make some simple and fundamental comparisons. Inspirations from biology that could have an impact on technology ("bionics") may also be drawn; however, they won't be discussed in any detail in the present book.

Photographing beyond a biological perspective

In this book as well, all of the photographs have been taken by Georg Glaeser (who also designed the book's layout). His approach to photographing animals combines empathy with artistic (and sometimes also mathematical) enthusiasm, rather than following the criterion of making pictures that are zoologically as easily identifiable as possible. The flight of the scarab on the left page or that of the barn swallow on this page are not just interesting in terms of their flight technicalities (the picture of the scarab shows how the beetle's wings function with the elytra closed). Additional information is provided by shadows and reflections on the surface. The numerous pictures that the photographer has taken of scarabs suggest that, before they rise into the air, these beetles tend to rotate in such a way that the sun shines on their backs, thus casting a symmetrical shadow. This habit probably helps the beetle to navigate using orientation cues from the sky, and the Ancient Egyptians, who associated the scarab with their sun god, surely noticed this as well.

Some animals had to be drawn

Flying animals, like all others, have developed over the course of millions of years. In the Carboniferous period, dragonflies whirred through the air. Birds evolved during the Jurassic period. Pterosaurs first appeared even earlier in the Triassic period and became extinct towards the end of the Cretaceous period. In the present day, many of

Seba's short-tailed bat (*Carollia perspicillata*)

Blister beetle (*Mylabris oculata*)

these animals can only be reconstructed and "drawn" – a task undertaken by the artist Markus Roskar. He also created illustrations of animals such as flying snakes and gliding ants, which could not be captured in high quality by camera.

Sometimes "multiple images"

Flying is a very dynamic process. Individual snapshots, while important and sometimes very spectacular, are often not as useful for the analysis of complex motion processes as "animations", which typically consist of a series of photographs taken and displayed in rapid succession (and sometimes films in super-slow motion). Insofar as technically possible, photographs may also be merged to illustrate motion processes – as was done, for instance, in the picture on the left-hand side.

Approaching a topic from different perspectives

This book may be read as a survey of flying animals across the evolutionary tree. In addition and alternatively to that, the authors would like to shed light on the evolution of flying from different perspectives and to draw comparisons between the different results of this evolutionary process. This shall primarily be done by means of photographs, which are occasionally complemented by schematic drawings. Since the focus will be on the photographs, the texts are concise while still including essential information.

Bibliography and web links

Bibliographical references to relevant literature and websites allow the reader to explore the topics that are covered in the book in more detail. Since online content is subject to change and may be deleted, the book is accompanied by a website, which keeps track of such changes.

Double-page principle

This book need not be read from front to back, as it is designed based on the double-page principle. Occasionally, cross-references are made to other pages. To facilitate the understanding of scientific terms when starting to read the book in the middle, it is recommended to consult the index to see where such terms are previously dealt with in the text.

An unbelievable variety

This book is meant to highlight and illustrate the various forms of flying that can be found in nature. For their contributions and help in creating this book, we would like to thank the following people (in alphabetical order and without academic titles): Daniel Abed-Navandi, Gudrun Maxam, Axel Schmid and Sophie Zahalka. Stefanie Wolf from our publisher Springer Spektrum has provided valuable and dedicated support during the planning and publication of the project. The limited number of pages required a very strict selection of photographs to be included in the book. However, numerous additional photographs can be found on the website that accompanies this book: www.uni-ak.ac.at/evolution

Contents

Chapter 1: 400 Million Years of Flight Evolution
Fourfold development to perfection   **1**

It must have been about 400 million years ago (mya): Certain insects started to develop wings, and not long after, almost all insects had gained the ability to fly. Birds evolved 200 mya, pterosaurians (from which birds do not have descended) probably even 230 mya. The fourth phase of this evolutionary process finally introduced flying mammals with the evolution of bats over 50 mya.

Chapter 2: Photographs of animals in flight
Challenging in many ways   **19**

This chapter will deal with several important aspects that need to be considered when creating exciting and insightful visual material. Photographs of flight are usually "action photographs" that require quick reaction. The wings of insects move so rapidly that they can only be "frozen" over fractions of milliseconds. In addition, flying animals are constantly changing their distance from the camera, which makes it more difficult to maintain focus.

Chapter 3: From the perspective of the biophysicist …
Fluids and scales · **51**

Flying occurs in the air, and in the figurative sense, it may also occur in the water. Both tiny and relatively large animals can fly. Simple parameters derived from biophysical studies of flight still allow useful comparisons. They provide a deeper understanding of the problems that animals encounter and solve while moving through air and water.

Chapter 4: Criteria of evolution
Sexual selection, climatic changes · **87**

One reason why evolution works so well is the existence of two sexes. These two sexes are joined in a variety of ways. In most cases, the female choses a male based on certain fitness criteria, which the male must demonstrate (female choice). Another way is, that males directly compete for the female through battle (male-male competition). If the next generation is supposed to fly better, then a major role is also played by mutations that permit the possibilities of biophysics to be stretched even closer to their limits.

Chapter 5: Insects: The first flying animals
Using every niche 123

Insects are the most successful class of invertebrates. Most of them have a more or less well-developed capacity for flight. This capacity proved to be such a major advantage in their evolution that it evolved relatively quickly and consistently across the class. Even giant insects like the African Goliath beetle and the stag beetle can fly.

Chapter 6: Birds: The "classics" among flying animals
From the hummingbird to the Andean condor 165

Among the vertebrates, the ubiquitous birds are the epitome of flying. The Archaeopteryx – considered a classic "missing link" in evolutionary theory – and its immediate ancestors may be taken as the starting point of all bird life 200 million years ago. Since then, birds have been flying, whirring, and gliding in all sizes, pollinating flowers or preying on insects and small mammals.

Chapter 7: Bats
Flying mammals **209**

It was only about 50 million years ago that bats and flying foxes took to the skies. Insects have been flying for a period seven times as long, and birds for a period four times as long. The pterosaurs, which had been flying through the air for some 170 million years, disappeared due to drastic climate changes towards the end of the Mesozoic Era. Navigation by means of ultrasound has enabled bats to conquer darkness.

Chapter 8: The fascination remains **225**
The topic is and will continue to be exciting

The fact that animals – and ultimately humans as well – can rise to the air as if gravity did not exist has always exerted enormous fascination on us. Biophysics has allowed us to understand and analyse this fact. Moreover, it is quite "typical" for the stunning capacity of evolution to use every imaginable niche and push it to its limits by placing, step by step, "one stone upon another."

A "classic": Otto Lilienthal's drawing of lift-generation with large birds. This topic, amongst others, will be explored in this book (cf. 60f).

1 400 Million Years of Flight Evolution

Fourfold development to perfection

It must have been about 400 million years ago (mya): Certain insects started to develop wings, and not long after, almost all insects had gained the ability to fly. Birds evolved 200 mya, pterosaurians (from which birds do not have descended) probably even 230 mya. The fourth phase of this evolutionary process finally introduced flying mammals with the evolution of bats over 50 mya.

Systems in the geological timescale

4.6 billion years, but 7/8of that time without fossilized traces of life

The Earth is 4600 million years old. Currently available methods make it possible to date the appearance of new life forms with relatively high precision and to establish models of periodization according to these findings. The first fossilized traces of life date back to 541 mya. Geologists and palaeontologists divide the interval that has elapsed since then into three eras.

Division of the "period with traces of life (541 million years)"

These three eras are known as the Palaeozoic (541-252,2 mya), the Mesozoic (252,2-66 mya) and the Cenozoic (66 mya to the present). These eras are further divided into systems: the Palaeozoic into six,

the Mesozoic into three, and the Cenozoic into three as well. Since these eras hold some significance for the evolution of life and will be referred to throughout the book, they shall be listed here with their time limits and some keywords as regards the appearance of typical life forms (those with wings printed in colour). The table follows the geological time scale that can be found on Wikipedia, but it is worth noting that there are several, slightly differing approaches and terminologies.

Palaeozoic (53.4%):

- Cambrian 541 - 485.4 – marine conchifera, worms, and algae
- Ordovician 485.4 - 443.4 – graptolites, trilobites
- Silurian 443.4 - 419.2 – placodermi, insects, first terrestrial plants
- Devonian 419.2 - 358.9 – ammonites, first apterygote insects, fish, tetrapods, tree ferns
- Carboniferous 358.9 - 298.9 – amphibians, first winged insects, ancient dragonflies, forests of club mosses and horsetails
- Permian 298.9 - 252.2 – amphibians, reptiles, coniferous trees

Mesozoic (34.4%):

- Triassic 252.2 - 201.3 – reptiles, dinosaurs, ichthyosaurs, pterosaurs
- Jurassic 201.2 - 145 – dinosaurs, early birds, early mammals, ferns
- Cretaceous 145 - 66 – marsupials, angiosperms

Cenozoic (12.2%):

- Paleogene 66 - 23.03 – flowering plants, primates, bats
- Neogene 23.03 - 2.588 – Hominidae
- Quaternary 2.588 - 0 – mammoths, ice ages, human beings

Illustration of these periods

Lets relate the "lifespan" of our planet to one single earth day. If the origin of Earth 4600 mya starts at 00:00, it is not until about 21:45 that the first winged insects appear. The first flying fish a bit later. The first birds and the last pterosaurs around 23:00. Gliding mammals may be found around 23:35. That is to say, the protagonists of this book have been living in the last $2\frac{1}{4}$ hours before midnight. (As a side note: According to this time scale, *Homo sapiens* do not appear until 3.6 seconds before midnight.)

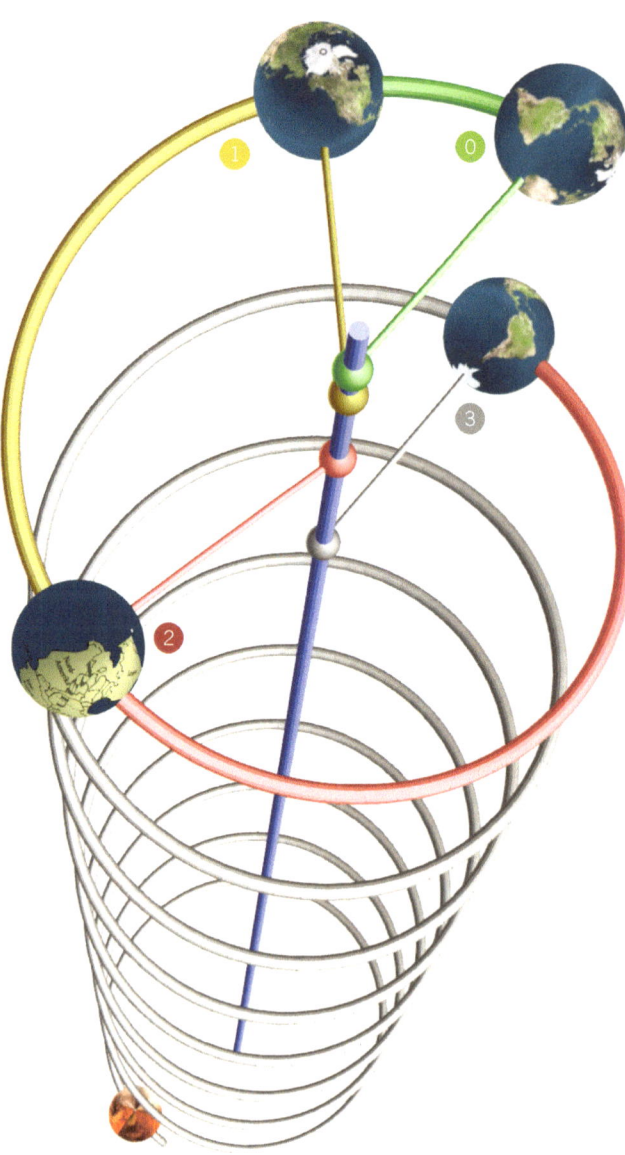

Only a tiny fraction of all animals survives

A billion animal and plant species are believed to have arisen (and largely gone extinct) at one time or another since the beginning of the Palaeozoic 541 mya. A much greater number of animal skeletons must have been "produced" in the process, which begs the question of what circumstances can prevent the complete decomposition of a living being's body after its death so that components, forms or structures of that body are retained: To date, more than a hundred thousand fossil species have been scientifically documented, but this represents only a tiny fraction of all animal remains that are "fossilized" (or using a broader term, "petrified", i.e. mineralized). Fossils are primarily found in sedimentary rocks (formed by the deposition of material at the surface of the Earth or under water).

When does "fossilization" succeed?

Bird skeletons, such as the ones pictured on this page, can survive intact or virtually intact for relatively long periods of time (at the top a bird skeleton when it was first found and on the right the same skeleton half a year later). However, in the long run, these skeletons will completely disappear as they are exposed to the effects of weathering. In this book, we will occasionally make reference to real fossils, such as the famous primeval bird *Archaeopteryx* (as an example of the "missing link", cf. p. 166) or relatively rare forms of organisms preserved in their original form in amber (cf. pp. see page 55f. and see page 41f.). Fossilization can only occur if the animal body is protected from further decomposition. So the dead remains must have been washed away to oxygen-free zones at the bottom of the Jurassic sea or they must have sunk into an oil-rich swamp (as it occurred to the numerous fossils found in the Messel pit near the German city of Darmstadt). The remains could then have been encased by showers of fine sedimentary lime particles or, at a later geological period, become embedded in oil shale, where they were eventually "petrified" with increasing pressure.

Replica of the "Berlin specimen" of *Archaeopteryx* (Belgrade)

Relative and absolute age dating

As an obvious "rule of thumb", the oldest layers of undisturbed sedimentary rocks can be found at the very bottom and the youngest layers at the top. Thus it is possible to categorize fossils in one layer in relation to those found in another layer. The long half-life of certain elements, such as uranium, thorium, and potassium, allows the absolute age of such rock layers to be determined.

Evolution – a constant process

Evolution – or: How do phylogenetic changes occur?

Evolution is defined as the change of organisms over time. It was recognized very early on that plants and animals had undergone a variety of such changes without knowing what had triggered these changes. It was recognized that offspring differed from their parents and that changes in any one generation are small compared to the differences observed across species. However, it was not until Darwin that the question of what triggered these large-scale changes between species could be answered. Darwin's theory of natural selection provides a causal explanation of how evolutionary change can occur.

Natural selection: Consequences for genetic composition

Natural selection is basically the difference in reproductive success when comparing two individuals from the same species. If this difference is not based on mere coincidence but on the fact that those individuals that are better suited to their environments produce more offspring than the other individuals of the same species. If this behaviour is heriditary, it will have consequences for the genetic composition of the next generation. So, this reproductive advantage, also known as 'fitness', must be hereditary.

Darwin's theses

In 1859, Darwin set out several theses regarding changes in organisms over time:

● Evolution means the changes that occur within populations of organisms over time. These changes are facts and not theories.

● These evolutionary changes occur in small steps. The extent of these steps can be correlated with the difference between parents and their offspring.

● A rise in the number of species results from the splitting of phylogenetic lineages, which occurs in addition to the evolutionary changes within those lineages.

● The mechanism underlying these phylogenetic changes is the process of natural selection, which Darwin defined as opposed to 'artificial selection' in the breeding of pets and agricultural crops.

● All organisms have descended from one common ancestor. The diversity of organisms is due to phylogenetic developments that have taken place over millions of years following the chemical evolution that has led to the origin of life. Hence, all organisms are related to one another.

The image shows a male *Neurothemis terminata*, a relatively large Southeast Asian dragonfly. Dragonflies already existed in the Carboniferous (especially "giant dragonflies" with a wing span of 72 cm) and they have barely changed since then. Not all dragonflies in the Carboniferous period were giants. Some of them were of "normal size". The giant size of the early dragonflies is sometimes believed to have been due to the higher levels of oxygen in the air (this would also apply to other insects during that period). However, some 150 mya ago, dragonflies seem to have become smaller in size without any significant decrease in oxygen levels. An alternative explanation could be that, with the appearance of the first birds, smaller species of dragonflies began to have an advantage as their larger and slower peers fell prey to birds.

All life comes from the sea

The Galápagos Islands have a strong historical connection with Charles Darwin. This photograph shows the famous rock formation known as "Darwin's Arch", situated near Wolf Island.

Below you can see the aquatic life that resides in these waters: a group of eagle rays (*Aetobatus Narinari*) in the foreground, a great hammerhead (*Sphyrna mokarran*) in the back.

All life on this planet began in the sea. The oldest known vertebrates date from the early Ordovician of about 450-470 million years ago. Cartilaginous fishes (rays, sharks) first appeared during the transition from Silurian to Devonian some 420 million years ago. It is no coincidence that the sight of swimming rays may trigger associations with flying. Leonardo da Vinci once said: "Observe the swimming of the fish in the water and you can understand the flight of the bird in the air."

Palaeontological evidence

During Darwin's lifetime, evidence to support his theses could be already obtained from palaeontology. Layers of rock serve as windows of time from the past, which can be precisely dated by means of modern methods. Clearly, only the most primitive organisms can be found in the older (earlier) layers of rock. Remains of the earliest vertebrates appear in later (younger) layers and birds and mammals even later.

Vertebrates at a later point in evolution

Even during Darwin's lifetime, these rock layers served as evidence to support that vertebrates had evolved after invertebrates, that birds had not appeared until the different groups of dinosaurs had already established themselves, and that such datable fossils could be used as conclusive evidence of evolution.

The first "missing link"

The discovery of the first primeval bird (*Archaeopteryx*) came in quite handy, as Darwin claimed that there had to be so-called "missing links" for his theses about evolution and natural selection, according to which major animal groups did not evolve independently but are united by connecting links, to hold true. Numerous such connecting links have since been discovered and can thus no longer said to be "missing links".

Scientifically verifiable opposition

Darwin's theses were and continue to be a scientifically verifiable opposition to creation myths, such as the Biblical Genesis and numerous other myths from around the globe, which hold that all species are the product of an individual creation event and have remained constant ever since. These claims have sparked a debate that continues to this day and has relatively little to do with science.

Establishment of the theory of natural selection

Breeders of animals and plants during Darwin's lifetime already knew that changes in traits could be attained by means of selective breeding. They would breed until, by chance, they managed to produce individuals that displayed the desire traits. That is, breeders would wait fo so-called "hot spots" to appear among their "samples" and then use these "hot spots" with the desired traits for further breeding. It is now known that these "hot spots" were actually mutations, i.e. heritable changes in an organism's reproductive cells. Through ongoing selection from one breeding cycle to the next, breeders gradually managed to attain their desired traits. Darwin assumed that similar mechanisms could be found in nature.

The *Archaeopteryx* constitutes a significant connecting link in the evolutionary lineage from feathered dinosaurs to present-day birds. However, as long as these connecting links are not known, that is, as long as they are still missing links, we can do no more than formulate hypotheses. The discovery of the *Archaeopteryx* thus not only revealed a prominent connecting link, but has also given support to the idea that birds have evolved from dinosaurs.

Differences in reproductive success

Since Darwin, natural selection has explained the difference in reproductive success that can be observed among individuals of a population and that depends on their genetic quality or fitness. This reproductive difference is, hence, no mere coincidence, but the consequence of these individuals' genetic constitution, and it will also have an effect on the genetic structure of a natural population's future generations. Those individuals that produce more offspring contribute to increasing the number of individuals that carry the more suitable genetic material. Since populations tend to remain constant in size, the proliferation of one kind of individuals necessarily implies the gradual disappearance of others that produce fewer offspring. That is to say, successful mutants gradually replace the less successful ones.

"Survival of the fittest"

Darwin referred to this as the "survival of the fittest" or "struggle for life". This survival of the fittest is not about animals fighting tooth and nail for survival, but a competition for reproductive success. So, rather than actual fighting, there is a competitive trial of strength. The winners are, for instance, those individuals that could run away from a cheetah as opposed to those that could not. The same principle applies to the cheetah, of course. Those cheetah that are faster than their prey are more successful in raising their offspring. Put very simply, this means that only the winners or the group of winners can successfully reproduce. It is usually enough for this group to produce more offspring than the others, and among their offspring it is once again only the fastest that survive as winners.

Selection based on environmental factors

The process of selection that is artificially guided during breeding by human beings is carried out in nature through the environment. This natural environment consists of so-called biotic and abiotic factors. Abiotic factors include, for instance, temperature and humidity. Biotic factors are based on the relationship between organisms living in this environment (particularly, their competition for resources).

Selection only occurs with genetic variation

Natural selection can only occur or be effective if the individuals of a population are not the same and differ in their genetic qualities. One 'trick' of nature that introduced such variation in the early history of living organisms was the invention of sexuality based on the division into two different genders that are joined in the reproductive process of meiosis (reduction division).

Rearranging the parental chromosomes

Meiosis is a special type of cell nucleus division, and it differs from the more common type of cell nucleus division, known as mitosis, in that the number of chromosomes is reduced by half. This implies a rearrangement, i.e. a new combination of the paternal chromosomes. The results of this type of cell division and the rearrangement

of the genetic material are the germ cells or gametes.

Each individual is genetically unique

Such cell nucleus division allows infinite possibilities of genetic variability. This shall be illustrated by means of a very simple example. Each individual is genetically unique. In order to see this, one only needs to look at us human beings. No individual is like another. Each individual has his or her own finger print. Thus in simple genetic terms, each individual is a separate and unique genotype. An example of a gene (1 locus) with one allele will yield 3 different genotypes among the offspring, which can be labelled as AA, Aa, and aa. Out of these three genotypes, two are homozygous (AA and aa) and one is heterozygous (Aa). A stands for dominant, a for recessive traits. So, for n genetic loci, there are 3^n genetically different individuals.

Exponential growth

That is, for $n = 20$ there are several billion genetically different individuals and for $n = 30$ more than 200 trillion. However, each individual consists of at least 1000 heterozygous genetic loci, which could yield 3^{1000} genetically different offspring. Since this number is so inconceivably great, it is reasonable to define the probability of two individuals being genetically identical as zero. Similar observations can be made with regard to the sorting of alleles. Two alleles per genetic locus yield 2^{1000} possible combinations of how these alleles can be arranged during reductive division, that is, before the production of sperm and ova. This number, too, is practically infinite. Each sperm, each ovum is genetically unique. All this taken together points to a source of practically infinite possible variations. If we take genetic mutations into account, the number of possible variations will increase even further. So, natural selection draws from a practically inexhaustible source of genetic diversity.

The opposite of chance, albeit a statistical process

The products of natural selection, that is, those individuals with more offspring than others within the same population, are not a matter of coincidence. It could be said that natural selection is actually the very opposite of coincidence. Still, natural selection is a statistical process, in which, as in a game of dice, individual cases cannot be said to be meaningful. In this process, merely the range of genetic variants is based on chance. It is impossible to predict which alleles will combine during meiosis and if and what kinds of mutations may occur. Such phenomena are simply a result of coincidence. However, which of these variants will eventually assert themselves in the process of reproduction is no longer a matter of chance. So, claims that human beings are the product of mere chance in evolutionary history are based on a misconception of what natural selection is and how it works.

When "the ability to fly" gives an advantage for survival …

When the possession of wings or organs for gliding gives an individual an advantage for survival, this will statistically lead to such individuals producing more offspring on average so that this trait is then further passed onto their descendants. If the evolutionary pressure to escape a predator is high, the possession of such organs is likely to assert itself quickly within a population, and it will also contribute towards the gradual improvement of these wings or gliding organs.

Flight organs determined by "parameters"

What kind of flight organs will evolve in this process is determined by various conditions and, of course, by the type of environment in which the organism in question lives. Physical laws provide a conditional framework channelling the evolution of wings and gliding organs, which are always subject to the laws of physics. Already existing limbs may also act as organismal preconditions in the evolution of wings, which are then adapted from these pre-existing structures.

The convergent evolution of wings in three vertebrates

This is what happened in the evolution of wings from pre-existing front limbs among vertebrates. Three groups of vertebrates convergently evolved such flight organs: pterosaurs, birds, and bats. Their

have a whole different origin, and how these wings may have evolved is still up for discussion.

Gliding mammals existed long before bats

Gliding organs have also evolved in multiple ways independently from one another. Well-known examples are the approximately 37 species of flying squirrels (also known as Pteromyini; members of the family Sciuridae, to which regular squirrels also belong), with most species occurring in Southeast Asia and the Himalaya region. Two species can be found in South America and there is even one (*Pteromys volans*) that is common from Northern Europe to Japan. Their size ranges from 7 cm (*Petinomys* and *Petaurillus emiliae* from North Borneo) to almost 60 cm (*Petaurista petaurista* from Southeast Asia), not counting the length of the tail. What they all have in common is a membrane of skin that attaches from the front legs to the hind legs and can be expanded like a sail. This extension of skin, which acts like a paraglider, allows them to jump from one tree branch to the next and to glide up to 100 metres, with their long tails providing stability during flight. Also worth mentioning is the insectivorous mammal (*Volaticotherium antiquus*), an ancient gliding mammal that actually bears no relation to the flying squirrel. This gliding mammal lived some 125 million years ago and reached a size of 12-14 cm. So this ancient animal could already be found flying across the air some 70 million years before the appearance of the first bats, but, like present-day flying squirrels, it glided rather than flew.

wings, though very different in their respective structures, gave these three animals the ability to fly. The flight organs of winged insects, also known as Pterygota (a term derived from the Greek word for wing),

Nachtigall W. **Gleitverhalten, Flugsteuerung und Auftriebseffekte bei Flugbeutlern**
In: Nachtigall, W. (Hrsg.): Biona-report 5. Bat flight - Fledermausflug. 171 - 186.
Akad. Wiss. Lit. Mainz, Fischer, Stuttgart (1986)

Rodents or marsupials – a convergent evolution towards gliding

Flying squirrels from the genus *Pteromys* and wrist-winged gliders from the genus *Petaurus* both glide by stretching the skin membrane between their front and hind legs. They look much alike, and with up to 20 cm body height and tails of similar length they are roughly the same size. Yet they are not directly related to each other, one being a rodent and the other a marsupial. According to studies, these animals can reach a gliding distance of at least 1.5 m with 1 m of height loss. So, their glide ratio would be no more than 1.5. However, from television films, we have also gathered evidence for significantly higher (better) gliding ratios. These animals can regulate their gliding flight by changing the curvature of the gliding membrane or moving their tails, which allows them to jump from one tree trunk, land on another and then climb up again.

Gliding reptiles

The flying dragons or gliding lizards from the genus *Draco*, with around 42 species distributed across Southeast Asia, are also gliders. They are members of the family Agamidae (lizards). They can reach a length of over 20 cm, with their thin tail accounting for more than half of this length. Males of the genus have a slightly serrated ridge on the back and a large, orange throat fan or gular flap. Females have a much smaller and blue flap. These lizards have a fold of membrane, brightly coloured with orange and black stripes, on either side of their bodies, attached to their free ribs, and they can expand these membranes to glide from tree to tree. They will also extend their gliding membranes when they feel threatened. The average distance

of their gliding flights is between 20 and 30 m, with 5-8 m of height loss. The maximum distance, however, can be up to 60 m. By twisting its tail, the flying draco lizard can control the stability of its flight. The movement of its tail and gliding membrane also allows the lizard to target specific destinations and avoid obstacles midglide. The fossil of the Cretaceous iguanians lizard, *Xianglong zhaoi*, in China, which is also believed to have used membranes stretched across elongated ribs for gliding, points to another example of convergent evolution.

Other gliding reptiles ...

A different form of gliding has been developed by 8 species of flying geckos from the genus *Ptychozoon*, which are endemic to Southeast Asia. Flying geckos are lizards with flattened bodies that may reach a length of 20 cm. Their tail is almost as long as their trunk. These reptiles are characterized by flaps of skin that grow from their flanks, head, tail, and limbs, as well as webbing between their toes. These skin membranes serve as sails and allow these lizards to travel short distances by gliding. Flying geckos even have the capacity to change direction during flight. They also have broad flattened toe pads equipped with adhesive lamellae.

Flying frogs

Some frog species from the genus *Rhacophorus* found across Southeast Asia have also developed something akin to flying membranes between their toes. Wallace's flying frog (*Rhacophorus nigropalmatus*) in Southeast Asia is named for the British biologist Alfred Russel Wallace. He was a contemporary and rival of Charles Darwin, and in the mid-19th century, he explored the Malay Archipelago, from where he brought the first flying frogs to Europe.

Gliding flights from tree to tree

The membranes between their toes allow flying frogs to glide distances of up to 20 m. Among the numerous species of flying frogs from the genus *Rhacophorus*, the best known is probably *Rhacophorus reinwardtii* (variously known as black-webbed tree frog, green flying frog, or Reinwardt's tree frog), which grows no larger than 8 cm. Flying frogs have extensive webbing or flying membranes between their long fingers and toes, which enables them to travel from tree to tree in short glides (see page 45f.). For further information on other strange kinds of flying animals, turn to pages f.

Dudley R.,Byrnes G., Yanoviak S.P., Borrell B., Brown R.F., McGuire J.A. **Gliding and the Functional Origins of Flight: Biomechanical Novelty or Necessity?** *Annual Review of Ecology, Evolution, and Systematics Vol. 38: 179-201 (2007)* Emerson S.B., Koehl, M.A.R. **The Interaction of Behavioral and Morphological Change in the Evolution of a Novel Locomotor Type: ``Flying'' Frogs** *Evolution 44 (8), 1931-1946 (1990)*

The largest flying animal of all time

Quetzalcoatlus – "the feathered serpent"

Nowadays, birds, insects, and bats (i.e. mammals) may be regarded as the rulers of the air. In the Jurassic period, however, ancient birds such as the famous *Archaeopteryx* and similar species were only starting to fly. Back then, and well into the Cretaceous period, the air was dominated by reptiles, more specifically by giant pterosaurs, among those the *Quetzalcoatlus northropi*, which, with a wing span as large as 12 m, is considered the largest flying animal to have ever lived.

Jumping upwards from cliffs, standing helpless on plains?

It was assumed that the biggest pterosaurs could only launch themselves by jumping from cliffs. So, if one of them had landed on a plain, it would not have been able to fly up again. However, more recent studies show that these pterosaurs could use their front wings as a pair of front legs. Compared to the wings of birds, the limbs of these pterosaurs were relatively strong.

Or did they launch themselves off "by jumping"?

It is more likely that pterosaurs launched themselves into flight by pushing themselves off the ground *with their four legs*, similar to what flies do with their middle and hind pairs of legs. They must have jumped to great heights to prevent their wings with the delicate membranes from hitting the ground when they started to flap.

Zakaria M.Y., Taha H.E., Hajj, M.R. **Design Optimization of Flapping Ornithopters: The Pterosaur Replica in Forward Flight** *doi: 10.2514/1.C033154 Engineering Science and Mechanics, Virginia Tech, Blacksburg, Virginia (2015)*

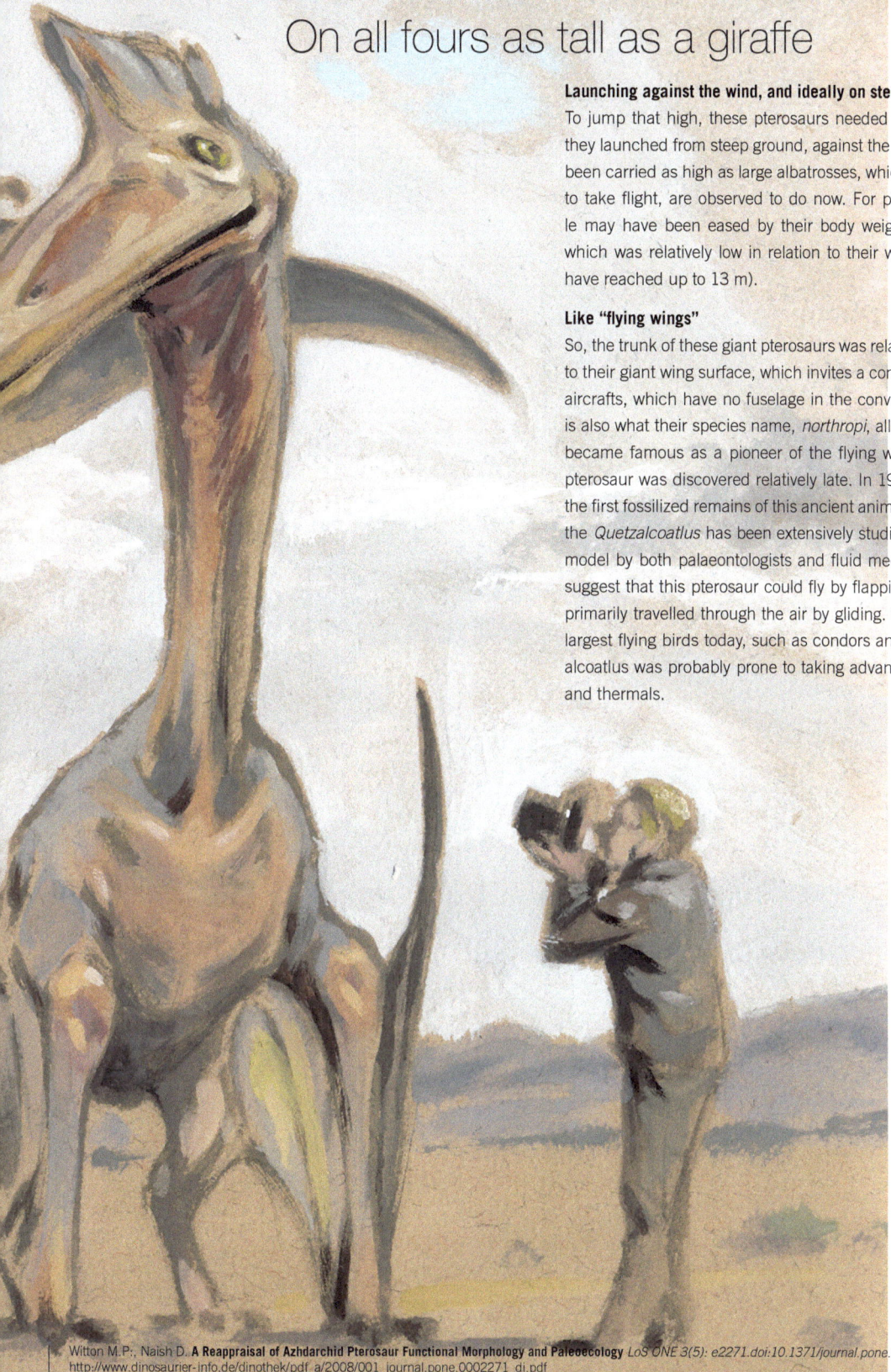

Launching against the wind, and ideally on steep ground

To jump that high, these pterosaurs needed a strong headwind. If they launched from steep ground, against the wind, they might have been carried as high as large albatrosses, which sometimes struggle to take flight, are observed to do now. For pterosaurs, this struggle may have been eased by their body weight of around 100 kg, which was relatively low in relation to their wing span (believed to have reached up to 13 m).

Like "flying wings"

So, the trunk of these giant pterosaurs was relatively small compared to their giant wing surface, which invites a comparison to flying wing aircrafts, which have no fuselage in the conventional sense. Which is also what their species name, *northropi*, alludes to: J.K. Northrop became famous as a pioneer of the flying wing design. The giant pterosaur was discovered relatively late. In 1971, a student dug up the first fossilized remains of this ancient animal's wings. Since then, the *Quetzalcoatlus* has been extensively studied and recreated as a model by both palaeontologists and fluid mechanics. Their studies suggest that this pterosaur could fly by flapping its wing, but that it primarily travelled through the air by gliding. Similar to some of the largest flying birds today, such as condors and vultures, the Quetzalcoatlus was probably prone to taking advantage of upslope winds and thermals.

Witton M.P., Naish D. **A Reappraisal of Azhdarchid Pterosaur Functional Morphology and Paleoecology** LoS ONE 3(5): e2271.doi:10.1371/journal.pone.0002271 (2008)
http://www.dinosaurier-info.de/dinothek/pdf_a/2008/001_journal.pone.0002271_di.pdf

Flying Fish

Flying fish

Also worth mentioning as a conclusion to this chapter are "flying fish". The most famous species of flying fish is the *Exocoetus volitans*, reaching a length of about 25 cm. Powerful tail strokes allow flying fish to jump out of the water. They then extend their elongated pectoral fins, with which they are capable of gliding up to several dozen metres above the water surface. They reach "heights" ranging from 1 to 10 m. Flying fish can thus escape many of their predators.

Hertel, H. **Biologie und Technik. Struktur-Form-Bewegung** *Krausskopf, Mainz (1963)*

Collective flight behaviour

Many fish try to escape their enemies by jumping out of the water. This looks particularly spectacular when it is done by a whole shoal of fish as it is chased, for instance, by a large predator.

The small hatchetfish (*Carnegiella strigata*, Gasteropelecidae) from South America can flap their long pectoral fins with such power and speed that they create a whirring noise while flying over water. They leap some centimetres out of the water and can fly distances of several metres.

Convergence

Amazilia hummingbird (*Amazilia amazilia*)

Dissimilar origins + similar selective conditions

In biology, the term 'convergence' denotes an evolutionary process in which similar structures or behaviour patterns arise from dissimilar origins due to similar or even identical selective conditions. Such structures are then referred to as analogous or analogies. The opposite of this would be homologies. Homologies are structures or patterns that share a common evolutionary origin, and their similarities are derived from this common ancestral form.

The wings of birds and bats are convergent

The wings of birds and bats are analogous because the last common ancestor that birds and present-day bats shared had not developed wings yet. So, the two kinds of wings must have evolved independent-

ly by convergence. In both cases, the front limbs, which originally served as pairs of legs, modified into wings.

From "normal" flight to hovering flight

The hovering flight of hummingbirds and that of the diurnal hummingbird hawk moth (*Macroglossum stellatarum*) also evolved independently. Both the hummingbird and the hummingbird moth derived from ancestors that could fly, but they had not evolved their extremely energy-consuming form of hovering yet. As its name suggests, the hummingbird hawk moth is often mistaken for a hummingbird by laypersons. However, it is unlikely that these two flying animals will ever meet, as one lives in the New World and the other one in the Old World.

Bird or insect?

Hummingbird hawk moth (*Macroglossum stellatarum*)

Convergence towards long-tongued bats

The hummingbird hawk moth is a member of the family *Sphingidae*, which consists mostly of nocturnal species. All members of this moth family developed their rapid, sustained flight ability to facilitate their search for flowers. It allows these animals to target specific flowers and to visit many flowers in a short period of time. It is also easier for these moths to fly away from flowers quickly in case of imminent danger. A further example of convergence is the hovering flight of the long-tongued Glossophaginae bats from the Southamerican family of leaf-nosed bats (Phyllostomidae). These bats are also known for being flower specialists.

Flying in the same Reynolds number range

The smallest hummingbirds and hummingbird moths may be compared with horizontal propellers, at least in terms of size, weight, wing shape (no curvature), wing surface and thus also wingload, wing movement, and the principle of lift to rise in the air. It could also be said that they fly in the same range of Reynolds numbers (cf. pp. see page 56f.). Conversely: Within the specified Reynolds number range, certain physical conditions need to be provided for an optimal hovering flight. These conditions, which match the ones listed in the first sentence, have been met by two very distinct biological lineages. The biologist would consider this as a case of convergent evolutions and analogous structures. In technology, this would qualify as an example of the phrase "form follows function".

Sprayberry J.D.H., Daniel T.L. **Flower tracking in hawk moths: behavior and energetics** (Journal of Experimental Biology 2007 Jan;210(Pt 1):37-45) 2007

2 Photographs of animals in flight

Challenging in many ways

This chapter will deal with several important aspects that need to be considered when creating exciting and insightful visual material. Photographs of flight are usually "action photographs" that require quick reaction. The wings of insects move so rapidly that they can only be "frozen" over fractions of milliseconds. In addition, flying animals are constantly changing their distance from the camera, which makes it more difficult to maintain focus.

Not clumsy at all

Almost like "dinosaurs"

This image of a brown morph of the red-footed booby (*Sula sula*) as it takes to flight might look scary. Its beak is, in fact, an effective weapon. However, the "story behind this photograph" goes as follows:

Trusting

This bird offered the photographer an impressive and touching performance on board a dive boat: After 10 minutes of "mutual communication" from a short distance (50 cm – the minimum distance of the camera lens) the bird flew a final "lap of honour" and twisted for a noisy "farewell" before it left. Without a doubt, the bird had established an empathic connection with the photographer, who took the decision on this same day to include the bird "in this book".

Possibly record-breaking

Common swifts *Apus apus* are among the fastest flying animals on this planet. They superficially resemble swallows even though they are not close relatives. These resemblances are due to convergent evolution. Common swifts travel long distances during migration. During breeding season from early May to early August, they like to stay in Central Europe. They then fly to their winter habitat in Africa, usually south of the Equator.

Motion blur

The common swift reaches a speed of 5-14 m/s in gliding flight, 11-28 m/s in flapping flight, and 40-60 m/s (more than 200 km/h!) when it takes a playful dive. The birds shown in the photograph flew at about 20 m/s. That is to say, it takes them 1/2000 s to move 1 cm. So, even when captured with short exposure time, these swifts are likely to produce motion blur, unless the camera tries to "move along with them".

This common swift was hunting too close above the surface of the sea and landed on the water. This would have been its death sentence if it had not been for an eager photographer who witnessed the situation and saved the bird. A few minutes later, the rescued swift was back with its flock ...

"Seagull photography"

Great white shark (*Carcharodon carcharias*)

Seagulls as constant companions of ships

Passengers of pleasure boats often throw food into the water (to the photographer's delight), in order to attract birds. Right-hand side: A seagull throws itself relentlessly into the water to catch fish offal that had been thrown into it. The commotion, however, has attracted another guest: The seagull has caught the attention of a 3-meter-long great white shark.

Sharks like to hunt for seagulls on the surface

This is a phenomenon that has often been observed. Tiger sharks actually travel to the lagoons of the remote Hawaiian Islands at the exact same time when albatrosses are ready to fledge, in order to catch some of the chicks that land on the water. It takes some experience to grab their prey successfully, though, because the bow wave that forms when the predator moves through the water may push the prey away from the shark's jaws. When birds and sharks hunt for schools of sardines at the South African coast ("sardine run"), the birds are not at risk, as both hunters focus on the sardines.

http://en.wikipedia.org/wiki/Sardine_run

The bottom picture was taken only a fraction of a second before the picture at the top. The seagull becomes aware of the critical situation and gets off to a lightning start. The way the water rolls off the bird's body shows that it must have taken off vertically.

Fighting rivals …

In the land of a thousandth of a second

If one wishes to take photographs of flying insects, certain techniques need to be practised. One possible method of practice could be to keep taking photographs of relatively common animals as they are encountered. Wasps and flies might be suitable animals for such practice. As time passes and more photographs of these animals are taken, one will begin to discover actions captured by these photographs that one would not have noticed with the naked eye. A wasp beats its wings three hundred times a second. This rapid movement can be captured by using flash. However, quick series of photographs cannot be shot in such a way. An exposure time of 1/8000 of a second can "freeze" the wasp's wings, while its surroundings will appear normally exposed.

Solitary and eusocial wasps

The family Vespidae is divided into a group of wasp species that live solitary lives and feed their larvae on their own and a second group of social wasps that form small or large colonies. Among the latter are also some of those dreaded wasps that are common in various regions of the world, such as the European hornet.

Nachtigall, W., Nagel, R. **Im Reich der Tausendstel-Sekunde** *Gerstenberg, Hildesheim (1988)*

The vibrant cycle of the year

Wasp queens hibernate before starting a new colony in spring. In midsummer, this new colony will grow considerably in size as unfertilized eggs produce male offspring and well-fed female larvae develop into new queens. Male and female wasps then leave the nest and mate. In autumn or early winter, the whole colony dies off, leaving only the newly fertilized young queens to hibernate and survive.

"Variants" among paper wasps (*Polistes*)

Several females, usually sisters, group together to build a small nest. These nests are wide open and attached to a plant stem or similar structure. What ensues is a kind of battle over which one of the females will take over the role as queen and which will merely serve as workers in the new colony. Only one female can win and become the sole queen of a colony. And it is only this one queen that lays eggs in the colony. Once again, a sexual generation of male and female wasps will emerge by late summer to mate in bushes or more exposed places. Competition is fierce, with male wasps engaging in aerobatic battles over the possession of females. The successful mating of two European paper wasps (*Polistes dominula*) is shown on this page. The male wasp in the background goes empty-handed.

Only females sting!

Male wasps and bees never sting. The female's poisonous stinger is actually a modified egg-laying device.

Copula of the paper wasp *Polistes dominulus*

Repeatable experiments

Common green bottle fly (*Lucilia sericata*)

A premise of experimental physics

The aim of natural sciences (and of experimental physics in particular) is to provide exact descriptions and explanations of events. Ideally, these events should be reproducible under similar conditions. In zoology, this is a wish that cannot always be granted. Animals do not always behave as expected, and they can react in many possible ways to a specific set of circumstances.

The flight behaviour of flies

If one wishes to "'reproduce'" a sequence of flight, one may be faced with much difficulty when it comes to rare animals or animals that have the capacity to fly but rarely do so. To capture a flight in all its details, laboratory conditions are required. There are animals

that are common and that can frequently be seen flying, such as the common green bottle fly or the housefly, which are among the most common species of flies. With some skills, one may get such flies to exhibit comparable flight behaviour.

Lateral swerve and vertical take-off

The picture series at the bottom (250 frames per second, 1/8000 s exposure time) shows the flight of a housefly from left to right (the interval between individual frames was lengthened to avoid overlaps). The picture series at the top shows the vertical take-off of a common green bottle fly. One must imagine these photographs being stacked one on top of the other to recreate the vertical flight of the fly (500 frames per second, 1/8000 s).

Housefly (*Musca domestica*)

Nachtigall W. **Insects in flight - a glimpse behind the scenes in biophysical research** *George Allen & Unwin, London (1968)*

Combination of jump and flight

The great green bush-cricket (*Tettigonia viridissima*) – pictured here, a male – is as quick to react as flies. So it is relatively easy to get these common insects to reveal both informative and comparable flight patterns. The great green bush-cricket, as opposed to many other cricket species, opens its wings as it jumps. Its flights are relatively short, and they usually only serve the purpose of escaping immediate danger. The photographs on this page were shot at an interval of 1/250th of a second (1/4000 s exposure time).

Vertical take-off preferred

The above series of images shows a classic example of a vertical take-off. Once again, one must imagine these photographs stacked on top of each other (which is easily achieved in this case by aligning the cricket's "springboard". In the second series, the cricket jumps just below the camera, coming into focus only briefly. Such image series are more difficult to create than individual photographs, but they are often very insightful.

The third series shoes the cricket's typical nearly-vertical take-off as viewed from the side (here as well, one needs to imagine these photographs merged together!).

Actual flight phases are relatively short

The final series shows a cricket as it actually flies parallel to the ground. Crickets tend to travel only a few metres in this manner.

What slow motion shows ...

The phases of the first fortieth of a second

Take-off jump

The fly faces the spectator as it takes off and then rotates by 90° to fly off to the right. The first image to the left shows the fly touching the ground with all its legs. In the second image, the fly lifts its front legs (take-off signal) as it uses its middle and hind legs to catapult its body up into the air (take-off jump).

Legs placed in airflow

The fly's hind legs are then drawn towards its abdomen, looking backwards. By stretching its limbs in different manners, the fly can variously place its legs in the airflow produced by the flapping of its wings. This calls to mind the navigation system of early rockets, which consisted of graphite components that were adjusted to the jet stream for steering. The fly's stretching of its front and hind legs appears more erratic in comparison. If the insect wants to land, it will remain in this position.

Closing all gaps

If it wants to continue its flight, the fly will place its middle legs, facing backwards, between its thoracic plate and abdomen, and its front legs, facing forwards, between its front thorax and head. In this manner, "all gaps are closed," which reduces air resistance and thus saves metabolic energy. The pictures were taken at an interval of 1/250 s.

Full bath within a third of a second

The things you discover in a bird bath

The two image series of a European bee-eater printed above and on the right-hand page were shot within a third of a second: The fast and flexible bird does a somersault as it dives into the water. Before it rises back into the air, it spins on its longitudinal axis to shake off the water.

Reaction time correlates with absolute size

The reaction time of small animals is incredibly short, which is actually due to their small size: Their neural paths are thus considerably shorter (with human beings, it takes 1/30s for an impulse to travel from the toes to the brain).

... and once again for the camera!

Blending images to consolidate information

Image blending is a technique that is frequently used to illustrate biological motion sequences. Before the invention of digital photography, a similar effect for the representation of animal flight was achieved by mapping the intermediate frames of a sequence onto the same image (ideally, placed in the right position and with a specification of the time interval between the individual frames). The photographs in this book were shot using high-quality digital cameras. In order to capture the basilisk or "Jesus lizard" jumping into the "life-saving water" as shown from various perspectives on this page, the photographer had to take 12 images per second.

Three consecutive photographs were taken from nearly the same perspectives so that it was possible to superimpose these images onto each other. The result (bottom right image) shows the basilisk in three different positions (at an interval of 1/12th of a second). In many cases, such "consolidated shots" are more informative than the full sequence of individual shots because they allow for a direct comparison of relative values such as the jumping distance and the movement of the lizard's limbs. Problems arise, however, if very short intervals between frames are required to capture a motion sequence because the body positions of the individual frames will overlap.

Basilisk (*Basiliscus vittatus*)

Burnet companion moth (*Euclidia glyphica*)

Grass moth (*Agriphila tristella*)

Shield bug (*Pentatomidae*)

Bugle sawfly (*Athalia cordata*)

This book includes many example of such "information consolidation". Four examples are printed on this page: A diurnal owlet moth at the top left, a grass moth at the top right, a stink bug at the bottom left, and the small image above this text shows a bugle sawfly as it takes flight. All images were shot with a very short exposure time (1/2000 to 1/8000 s) and at an interval of 1/12 s.

The nightmare of high-speed photography

The so-called rolling shutter effect
is an effect that is well known in the film business and that can occur in photographs or videos of rapidly moving objects. Most cameras cannot expose the whole sensor surface to light at once. Instead, imagine exposure occuring in a line from one side of the image to the other, either horizontally or vertically – a process which takes a short though not inconsiderable amount of time.

s). The body of the beetle "barely" moves, while its elytra (wing cases) move significantly faster than its body (they flap during flight). The wings themselves, on the other hand, oscillate at a high speed relative to the beetle's body.

What every "high-speed photographer" needs to know
The shutter time of most camera models is usually considerably longer than indicated. Since the exposed sensor is scanned line by line

Red-brown Longhorn Beetle (*Stictoleptura rubra*)

Can you rely on one photograph?
The photograph on this page, which could almost qualify as art, shows what can happen when a flying longhorn beetle (*Stictoleptura rubra*) is photographed using a very short exposure time (1/25000

and each scanning of a line takes only a fraction of the shutter time, the exposure time for each line is reduced accordingly. However, it takes the device this same (very short) amount of time to move to the next line, which is once again exposed very briefly. During that period of time, the wings may have covered a certain distance, which then causes such effects as skews, wobbles and smears.

https://www.youtube.com/watch?v=CmjeCchGRQo

What goes unnoticed while watching a film …

Right-hand side: Shots from an HD-video (30 frames per second, exposure time was manually set to 1/2000 s). The faster the wings move, the more they appear blurred (though this is usually only noticed when looking at the individual frames).

When the shadows do not match

The shadows of the beetle's wings appear lower with every shot, which is due to the delay of a few milliseconds while scanning the many "lines" of the sensor. To recreate this effect with a technical device, several photographs were taken of a miniature helicopter (roughly the size of a giant dragonfly and with rapidly spinning rotor blades). Although the images were shot using an extremely short exposure time of 1/32.000 s, the rotor blades, which are actually symmetrical and not curved, appear distorted (as opposed to the labels on the blades, which can clearly be read)!

The problem lies in the closeness to reality: Image errors may look so realistic that one is tempted to take them at face value (consider, for instance, the second-to-last image of this series)! Especially in high-speed photography, supposedly realistic images must be taken with a pinch of salt! Certain tests should be carried out to check if such image errors are actually possible. Shadows provide the best starting point for such a reality check. Further instances of "rolling-shutter distortion" in this book (usually images of insect wings oscillating at maximum speed) will be pointed out!

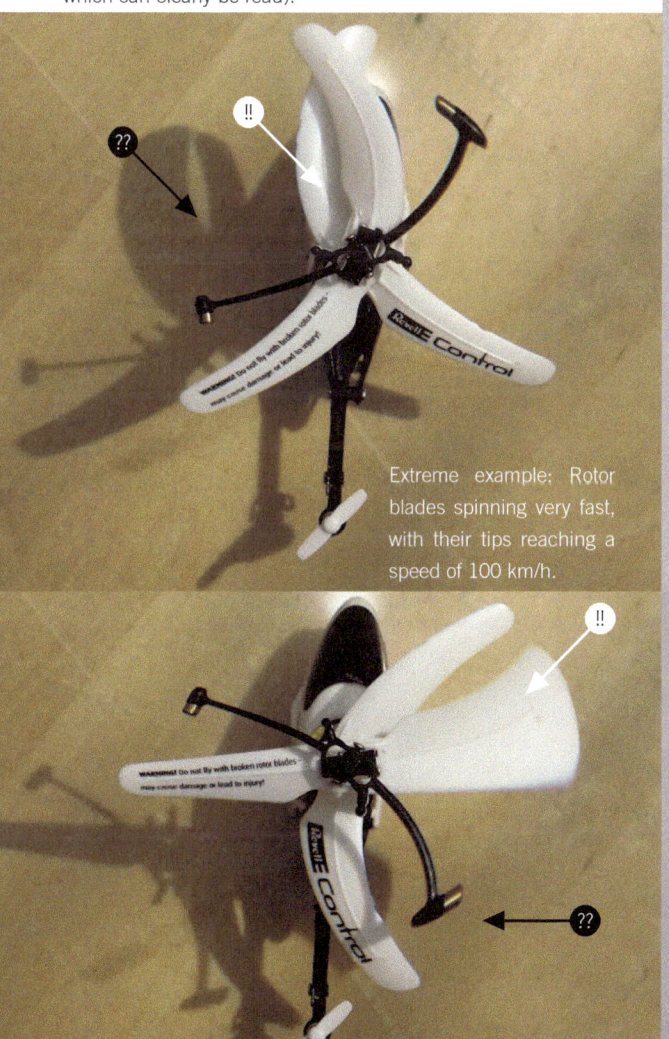

Extreme example: Rotor blades spinning very fast, with their tips reaching a speed of 100 km/h.

Female of the Red-brown Longhorn Beetle (*Stictoleptura rubra*)

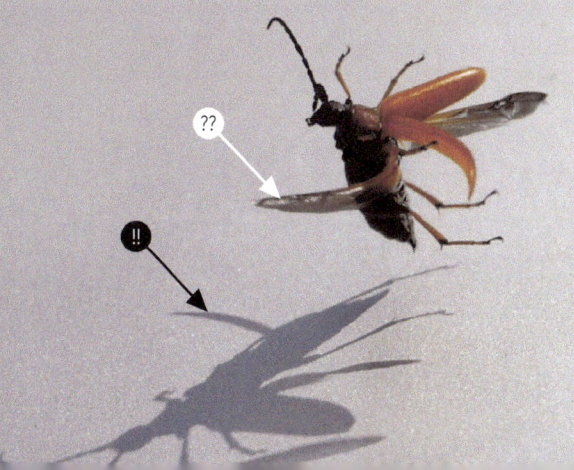

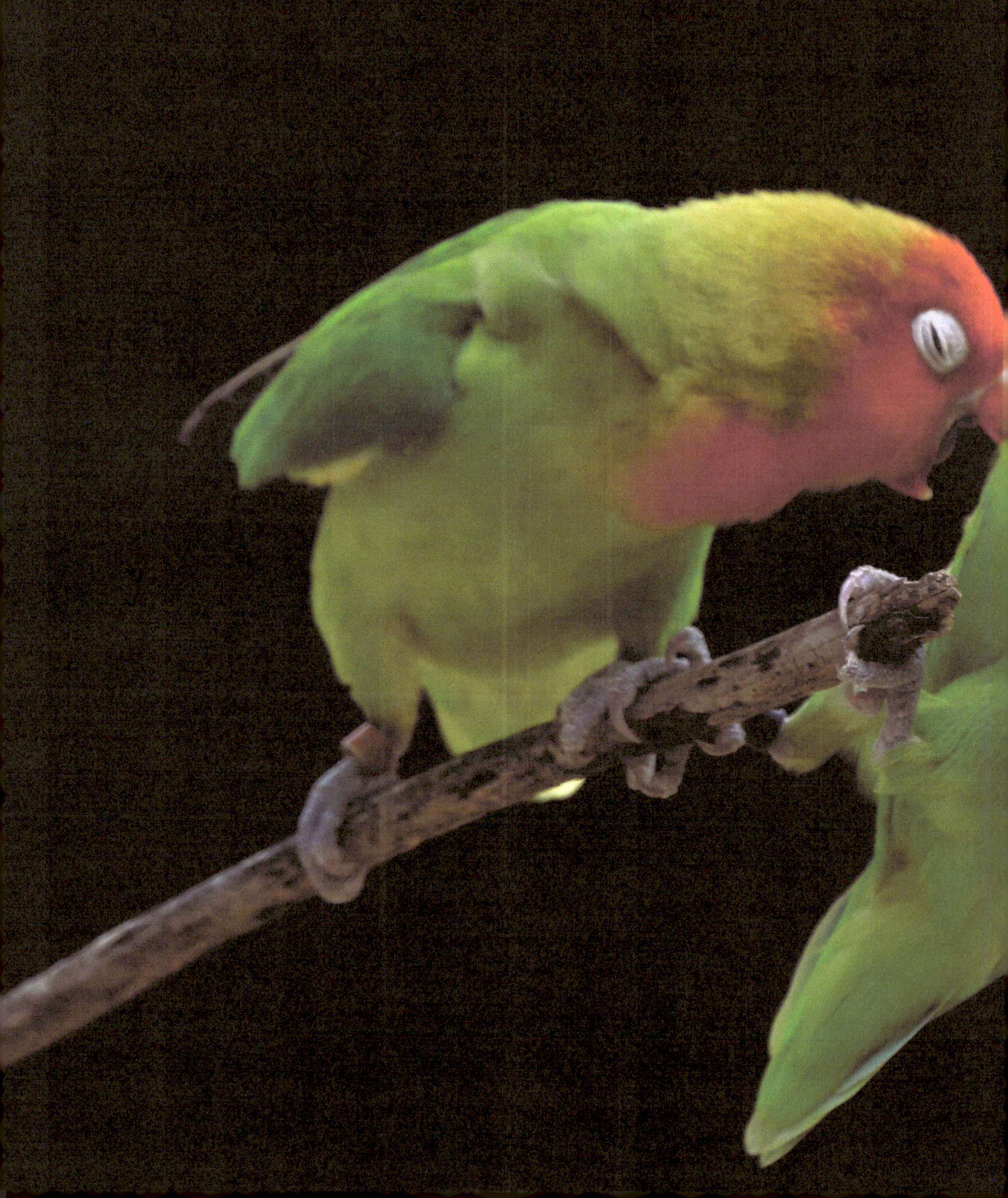

Male and female **scarcely differ.**
The parrots from the genus *Agapornis*, some of which are only 13 cm long, are commonly known as "lovebirds". They are monogamous and usually remain together throughout their lives. They come in a variety of colours, but the males and females scarcely differ from each other. They are also known for their strong bonding behaviour, as they show their affection by constantly preening and feeding each other. That is why they are also known as "the inseparables" in German. The couples like to breed in the caves or cavities of East Africa, occasionally taking over nests that were abandoned by weaver birds. Both parents raise and take care of their chicks, even weeks after the young birds have left their nest.

With claws and beaks Parrots are generally known for being outstanding climbers. For this task, they use both the claws of their feet and their flexible beak. Along with ravens, they are among the most intelligent birds.

True splendour

Male of Banded Demoiselle (*Calopteryx splendens*)

Male of Beautiful Demoiselle trying to mate with an egg-laying female
(*Calopteryx virgo*)

Large damselflies with gender-specific colouring

The metallic-shimmering damselfly (*Calopteryx*) is quite an eye-catcher with its wing span of 6-7 cm and body length of 5 cm. The males of the banded demoiselle have a shimmering dark-blue-green body and translucent wings with a greenish tinge which each have a broad, blue-black band. The females are less conspicuous with their metallic green-bronze colouring and their iridescent green wings. The males of the beautiful demoiselle also have a metallic shimmering body and iridescent dark-blue wings. Both species prefer slow-moving wide streams and rivers with clear water to develop their larvae.

Spectacular courtship flights of the males

The males defend their small territories against other males. When they meet a female, they will perform a spectacular courtship flight on the spot or right above the female. They flap their wings in alternating rhythm (one pair of wings goes up as the other goes down). After each flap, they hold their wings for a few tenths of a second before they flap them into the other direction. These movements are probably supposed to draw attention to the bands and colours of the male's wings, and they also allow the males to display the brightly coloured tip of their abdomen.

The famous heart-shaped mating wheel

The pair then flies together around the territory, looking for places to lay their eggs. Copulation begins once the female lands. The male then grasps the female by her prothorax with his anal appendages, and the female curls her abdomen to lock with the secondary genitalia at the base of the male's abdomen.

"Sperm competition" even during egg-laying

The male then releases the female and flies around her to guard her as she lays her eggs. Rival males often try to grasp a female that is still in the process of egg-laying. These rival males are immediately attacked by the female's mating partner. The two photographs on the left page show such a situation. Damselflies are also well-known for a particular mating strategy that was first observed with these insects: During copulation, the male scrapes at the female's abdomen in order to remove any sperm from a previous mating and replace it with its own sperm. This phenomenon of sexual selection is also known as sperm competition, and it illustrates very well how this competition to produce offspring includes taking measures to suppress the reproduction of rivals.

"Flap by flap"

Equal rights for both parents

The chicks of great tits, like those of Eurasian blue tits, are fed by both parents. Thus, they are often seen closely following each other as they approach and leave their nest box. While one is about to carry out droppings in order to leave them somewhat removed from the nest, the other is bringing new food in its mouth.

Spreading their wings

The present photograph clearly shows to what extent birds can spread the distal ends of their wings, on the one hand to pick up speed and on the other to pull the "air brakes". The breast stripe of the female (taking to flight) is narrower and more discontinuous than that of the male (landing), as are the black lines on the sides of the neck.

Not really migrant birds

Great tits are generally resident birds, which rarely move far away from their breeding territories. In Europe they can even be found throughout the year in the regions north of the Arctic Circle, where they typically benefit from human feeding during winter.

Mimicry

Several hoverfly species resemble wasps, bees, and bumblebees in their forms and patterns, but they have no stinger. By means of this mimicry, hoverflies trick their enemies into a false sense of danger. Their wings produce a figure-eight loop 300 times a second. The huge eyes are particularly beneficial to the males during their search for a partner.

With closed elytra

Fast take-off with a deep humming sound

Flower chafers (on this page: *Potosia cuprea*), like scarabs (see page 64f.), do not open their elytra during flight. Instead, the wings are stretched out sideways. This is possible due to notches at the lateral wall of the elytra, which are covered with bristles to give protection against sweeping dust. This provides certain advantages for take-off: Flower chafers can take flight extraordinarily fast, though their speed during flight is less extraordinary. They are beaten, however, by the tiger beetle, which can take flight even faster despite having to raise and fix its elytra like aerofoils before take-off. The heavy rose chafer produces a deep humming sound during flight. Its wingbeat frequency ranges from 70 to 130 beats per second, depending on the size of the beetle.

Not a bumblebee!

The large bee-fly (*Bombylius major*) typically visits flowers in hovering flight. In these photographs, the large bee-fly is stretching its legs as it is about to land on a flower. It does so to lift its body and make room for its very long "proboscis". The large bee-fly can also stay still in hovering flight. It often does so in front of the entrance of a mining bee nest, into which it will deposit its eggs, which are previously coated with sand for camouflage. The larvae hatching from these eggs will live parasitically among the young mining bees. The oldest fossil remains of this insect family are preserved in Siberian amber dating back to the Cretaceous period (70-100 mya).

Schremmer F. **Gezielter Abwurf getarnter Eier bei Wollschwebern (Dipt. Bombyliidae)** *Zool. Anzeiger 27: 291-303 (1964)*

The first step towards gliding flight

Basic requirement: Generating "airflow" when there's no wind

When a lizard or a frog jumps into the air, they have already fulfilled the first requirement for generating "aerodynamic force": They have produced airflow, that is, a stream of air coming from the front, even when there is no wind.

There is not much to be done from a horizontal plane …

If they jump from a horizontal plane, they will land on the ground again after a "short jump trajectory", and they will reach their highest speed right after the jump. Such short jumps may help these animals to escape a predator, but they will not get them far "aeronautically".

However, if it goes downhill ...

The situation is different, however, if they jump from a height, such as a tree. They will descend a certain distance until they hit ground, which allows them to pick up speed. The frog in this image jumped from a car roof. It thus covered a flight distance of three meters, which equals 60 times its body length! What is also noteworthy is the position of the frog's legs during its flight.

It only takes a few tricks to achieve gliding flight

Aerodynamics forces increase as speed rises (more precisely, by its square). If the animal has developed some kind of gliding surface that is inclined to the air stream at an angle, such as a membrane stretched between the ribs, as it is the case with flying dragons, or the webbed toes of flying frogs, then it is possible for these animals to generate lift force by means of these gliding surfaces and to increase the gliding distance. This may then be considered as a form of gliding flight. The first flight attempts by the primeval *Archaeopteryx* may have followed a similar strategy (cf. pp. (see page 166f.).).

Gliding and running across water

The "Jesus lizard" earned its name from its ability to run across water at speeds of about 8 km/h in order to escape its enemies. To do this, these lizards usually jump, or actually glide, from a branch into the water. Below: multiple shots (four shots with an interval of 1/12 second each).

A young black-webbed tree frog (*Rhacophorus reinwardtii*) spreads its front and hind limbs at a certain angle while jumping. It also spreads its toes, which are connected by webbing. Each foot thus acts as a kind of gliding surface, but due to the slight spreading of the limbs, the frog's trunk and feet function together as one unit. How exactly the partial airflows interact is not sufficiently understood yet, but they permit gliding at an angle of 25°. Their aerodynamic quality thus equals at least that of a beermat gliding in perfect balance (glide ratio of about 2.5).

Below: The flying frog jumps vertically into the air (both series: 250 shots per second).

Adaptable flight artists

Masters of the stormy winds

Common ravens (*Corvus corax*), along with the red-billed chough and the alpine chough, are the flight artists among the highland birds. They are masters when it comes to flying in stormy winds: They let themselves be lifted by the winds, take a nosedive into valleys, and never lose control in the process. They can even cope with the strong winds that are generated by the waves behind mountain ridges.

More than 100 km/h!

To minimize loss of altitude, common ravens try to cross these regions in fast gliding flight. For this purpose, they bend their wings to reduce the lifting surface and thus increase the wing loading (weight divided by the surface). This will steepen their gliding angle, which allows the common raven to pick up speed of more than 100 kilometres per hour. In steep gliding flight, the rear edges of the wings act as a type of manoeuvring flap system that allows them to control the airflow hitting the raven's tail. Common ravens thus have the capacity to navigate with great accuracy even at high speed, which comes handy in situations that require very precise manoeuvring skills, such as nose-dives into narrow ditches and canyons. They also make use of their peculiar navigation system during courtship, doing spectacular rolls, somersaults, and loops.

Küttner, J. Über die Flugtechnik einiger Hochgebirgsvögel *Kosmos 384 - 389 (1947)*

How do you open an oyster?

Oysters are a delicious meal, for both seagulls and baboons alike, but they are not easy to shuck. The kelp gulls (*Larus dominicanus*) at the Cape of Good Hope (South Africa) have devised a method to solve this problem: They take an oyster high in the air, up to a height of about 20 metres, and then drop it on a hard rock. However, after shucking the oyster on the rock, they have to act quickly, because baboons have learned from witnessing this spectacle and are fast on the scene.

Spectacular courtship

Frigatebirds (Fregatidae) are a family of tropical and subtropical seabirds from the order Suliformes. The males of this species are known for and recognized by their inflatable red gular pouch, which they use to attract females during the mating season. The red colouring of the gular pouch, the size of the inflated gular pouch and stamina are likely to be assessed by the female as indicative of the male's fitness. On this page: Male magnificent frigatebird (*Fregata magnificens*) in the Galapagos islands. At the bottom left, you can also spot a Galapagos land iguana (*Conolophus subcristatus*).

Egg thieves

Frigatebirds prey on the eggs that marine iguana (*Amblyrhynchus cristatus*) lay in warm sands, eager to steal several of these "energy bombs". The "scramble" for the eggs often starts in the air, and the frigatebird may lose some of its loot during the fight. The bird's name is actually derived from this "kleptoparasitic" behaviour (animals stealing food or prey from another), which is reminiscent of the frigate ships that pirates used to attack merchant vessels.

3 From the perspective of the biophysicist …

Fluids and scales

Flying occurs in the air, and in the figurative sense, it may also occur in the water. Both tiny and relatively large animals can fly. Simple parameters derived from biophysical studies of flight still allow useful comparisons. They provide a deeper understanding of the problems that animals encounter and solve while moving through air and water.

Records among flying animals

The values compiled on this page should not be considered absolute records, as new things are continuously being discovered. In any case, they may be regarded as coming close to the extremes.

● Wingspan

Largest wingspan among pterosaurs: Giant pterosaur *Quetzalcoatlus northropi* 10-13 m. Largest wingspan among living birds: Andean condor *Vultur gryphus*, large vultures, albatrosses about 3 m; smallest: bee hummingbird *Mellisuga helenae* from Cuba, wingspan 9 cm.

Largest wingspan among fossilized insects: Giant dragonfly *Meganeuropsis permiana* 70-72 cm.

Largest wingspan among butterflies: white witch moth *Thysania agrippina* from South America 32 cm, Atlas moth *Attacus atlas* from Southeast Asia, 30 cm, smallest: pigmy moth (Nepticulidae), 2-2.5 mm.

Largest wingspan among other insects (dragonflies, cockroaches, grasshoppers, cicadas): 17-23 cm, smallest: small thrips (*Thysanoptera*) and fairyfly (Mymaridae) ca. 1.5 mm.

Buff-tailed bumblebee (*Bombus terrestris*)

Flindt, R. **Biologie in Zahlen. 2. Aufl.** *Fischer, Stuttgart (1986)*

• Mass

Greatest mass among pterosaurs: giant pterosaur *Quetzalcoatlus northropi* over 100 kg.

Greatest mass among living flying birds (condor, vulture, great bustard): 12-18 kg, lowest mass (bee hummingbird *Mellisuga helenae*) 1.6 g.

Greatest mass among insects (15 cm-long larva of the Goliath beetle *Goliathus regius*) 110 g, lowest mass (thrips, fairyflies, fruit flies) 1 mg.

• Wing surfaces and wingloads

Largest wing surface among birds: griffon vulture *Gyps fulvus* 104 dm^2, smallest wing surface: ruby-throated hummingbird *Archilochus colubris* 0.1 dm^2.

Greatest wingload among birds: mute swan *Cygnus olor* 170 N/m^2, Smallest wingload: kinglet *Regulus spec.* 11 N/m^2.

Largest wing surface among insects (butterflies): white witch moth *Thysania agrippina* 400 cm^2, compared, for instance, with privet hawk moth *Sphinx ligustri* 26 cm^2, smallest wing surface among insects: chironomids (Chironomidae) 0.05 cm^2, smallest insects by 0.01 cm^2.

Greatest wingload among insects: bumblebees, dung beetle, water scavenger beetle 1-1.6 N/m^2, lowest wingload among insects: green lacewings Chrysopa spec. 0.05 N/m^2.

Greatest wingload among bats: slow flyers 7 N/m^2, fast flyers (Molossidae) bis 35 N/m^2.

• Wingbeat frequency

Lowest wingbeat frequency among birds: condors, large vultures about 1 Hz, greatest: white-vented violetear (*Colibri serrirostris*) 78 Hz.

Lowest wingbeat frequency among insects: desert locus *Schistocerca gregaria* ca. 20 Hz, greatest: bees, flies up to 300 Hz, smallest micetophilids (Mycetophilidae) possibly over 1000 Hz.

Wingbeat frequency among bats: straw-coloured fruit bat *Eidolon helvum* 7 Hz, lesser mouse-eared bat *Myotis blythii* 18 Hz.

• Speed

Greatest nose-dive speed among birds: peregrine 320-389 km/h, white-throated needletail *Hirundapus caudacutus* from Southeast Asia 335 km/h.

Greatest long-distance-flight speed among birds: geese, ducks, terns about 100 km/h, swifts about 150 km/h.

Greatest long-distance-flight speed among insects: deer botfly, pale giant horsefly, large dragonflies and sphingids 50-60 km/h (short-term).

Flight speed among bats: small bats (e.g. *Pipistrellus hesperus*) 10 km/h, large migratory species (e.g. *Tardaria brasiliensis*) 40 km/h (extreme value 60 km/h). A recently published paper (see below) reports that some small bats are able to fly even faster than the fastest birds.

• Metabolism

Lowest metabolic rate among birds (long-distance flight at 10 m/s): laughing gull *Leucophaeus atricilla*, Chihuahua raven *Corvus cryptoleucus* 10 ml O_2 per gram of body weight and hour, greatest metabolic rate: Mexican violetear *Colibri thalassinus* 60-70 ml O_2 per gram of bodyweight and hour.

• Migration distance

Longest "non-stop" -migration distance and flight duration among small birds: ruby-throated hummingbird *Archilochus colubris* (Gulf crossing) 800 km, 18 h. New World warbler 4300 km, up to 100 h. Latham's snipe 5000 km.

Migration distance of bats from summer to winter habitat usually no more than 60 km, but: hoary bat *Lasiurus cinereus* migrates from Canada to Florida: ca. 4000 km. Nathusius's pipistrelle *Pipistrellus nathusii* migrates, for instance, from Estonia to southern France: ca. 2000-2500 km, straw-coloured fruit bat *Eidolon helvum* from Zambia to Congo, ca. 3000 km (in huge swarms).

• Flight altitude

Maximum flight altitude among birds: Migration over the Himalaya by geese, ducks, and waders 7-9 km. Jet engine accident caused by Rüppell's vulture *Gyps rueppellii* at a height of 11.2 km (passively swept up by thermal currents?).

Nachtigall, W. **Der Flug der Fledermäuse** *In: Gaida K.G., Prokot S.:*
Microchiroptera. Falter, Wien (1992)
McCracken G., Safi K., Kunz T., Dechmann D., Swartz S., Wikelski M.
Airplane tracking documents the fastest flight speeds recorded for bats
Royal Society Open Science 3: 160398. doi:10.1098/rsos.160398
Max-Planck-Institut für Ornithologie,
http://rsos.royalsocietypublishing.org/content/3/11/160398 (2016)

Decisive physical conditions

The wings of animals perfectly reflect the possibilities that arise from the physical conditions of their surroundings. Unsuitable traits do not assert themselves in the process of evolution. Thus, it is of no little significance, whether wings flapping in the air are small or large. The air will have a completely different effect on small wings than on large ones. For small wings, the air becomes a relatively viscous medium, comparable to the consistency of light oil. Such physical conditions, therefore, require small wings to evolve with completely different structures. It is thus a mistake to believe that small bird wings are merely a geometrically shrunk version of larger wings.

Reynolds, O. **An experimental investigation of the circumstances which determine whether the motion of water shall be direct or sinuous, and of the law of resistance in parallel channels** *Phil. Trans. Roy. Soc. London 179, 935 - 982 (1883)*

The Reynolds number

The parameters that are relevant here are specified by the so-called Reynolds number, which is named after its originator:

$Re = v \cdot l \cdot \nu^{-1}$ (for air dwellers: v relative velocity (here: that of the respective wing part) to the surrounding air (m s^{-1}), l characteristic length (as regards wings: wing depth) (m), ν kinematic viscosity of air at 20° C equal to $1,51 \cdot 10^{-5}$ m s^{-2}).

The morphology of animal wings
and their aerodynamic characteristics

Re-effects are a decisive factor both in technology and biology; the morphology of animal wings and their aerodynamic characteristics – as reflected, for instance, by their Lilienthal polars – are in accordance with the respective range of Reynolds numbers. In phylogenetic history, species complexes are prone to "evolutionary radiation," that is, the opening up of new habitats not yet occupied by a certain species complex. For this purpose, morphological, physiological, and behavioural parameters evolve to achieve optimal adaptation to this new ecological niches. With regard to the parameters of the Reynolds number, one could call this process the colonization of Reynolds niches.

● The wing size can change

One agent of evolutionary change is the body size, and thus also the wing size, of flying animals, which may grow or diminish. The insect world, for instance, has witnessed an evolution towards ever smaller flying animals (see page 56f.).

● The wing morphology can change

Small animals typically move at a slower speed. So, the Reynolds number decreases with body size, as both the parameter l and the parameter v are reduced. For fluid-mechanical reasons, smaller wings generate less lift (lift coefficient c_A), but the ability to generate resistance (drag coefficient c_W) increases. This represents also a decrease in the lift-resistance ratio $\varepsilon = c_A/c_W$, which is also known as a lift-to-drag ratio. This, in turn, implies a change in the wings' morphological structure because the optimal settlement of different Re ranges requires different optimal wing structures.

● Evolution is bound by physical laws

All sorts of biological variations arise from the trial-and-error nature of the mutation-selection process. However, only the ones that follow the laws of physics will assert themselves as optimal. Profiled wings, for instance, are optimal with high Re numbers.
With low Re numbers, on the other hand, wings that are profiled are less efficient than wings that are not. So, in the course of evolution towards optimal forms, small wings lose their profiling. Even smaller numbers lead to a loss of curvature, and with the smallest numbers,

the wing blades eventually dissolve into bristles. This can be nicely illustrated by means of four examples of flying animals of different sizes. Listed next to these examples, you will find some characteristic details and approximate average values.

Four examples:

● Bird: l = 0.1 m; v = 15 m/s; $Re \approx 10^5$;
profiled and curved wings; $\varepsilon_{opt} \approx 8$
● Swallowtail: l = 0.04 m; v = 3.5 m/s; $Re \approx 10^4$;
non-profiled but curved wings; $\varepsilon_{opt} \approx 2.5$
● Blow fly: l = 0,0035 m; v = 4 m/s; $Re \approx 10^3$;
non-curved, flat wing; $\varepsilon_{opt} \approx 1.5$
● Insects with bristle-wings: l (bristle thickness) = 0.01 mm;
v = 1.5 m/s; $Re \approx 10^0$;
bristled wings; $\varepsilon_{opt} \approx 0.25$

Nachtigall, W. **Some aspects of Reynolds number effects in animals** *Math. Meth. in the applied sciences 24, 1401 – 1408 (2001)*
Nachtigall, W., with the collaboration of A. Wisser **Ökophysik. Plaudereien über das Leben auf dem Land, im Wasser und in der Luft** *Springer, Berlin etc. (2006)*

Movement in fluids

In air and water

Both air and water are fluids, and from a fluid-mechanical perspective, they can thus be treated equally. During flight, animals need to compensate for their body weight by generating lift, as in this image of a Eurasian blue tit as it is about to land. The bird's landing flight will later be discussed in more detail (cf. pp. see page 80f.). With its wingbeats, the Eurasian blue tit must generate both lift and thrust (or, as in this case, brake force), which reduces flight performance. In water, animals can compensate for their body weight passively, for instance, by means of fat deposits or swim bladders. Water fleas almost fully compensate for their body weight, which is why they only sink down slowly. With a powerful beat of their rowing-antennae, they manage to generate sufficient lift to rise after letting themselves briefly sink into the water and get back to their initial level, but they can focus most of their energy into generating thrust. In this respect, water dwellers have it easier than air dwellers.

The correlation via the dimensionless Reynolds number

$Re = \frac{l \cdot v}{\nu}$ describes the quotient of inertial force to viscous force, which act on a body in a fluid. This number increases in proportion to both the object's size l and the flow velocity v, and it decreases (indirectly proportionally) with the kinematic viscosity ν of the fluid. The latter depends to a large extent on the temperature, and at room temperature, for instance, it is 15 times greater in air than in water. That is why objects moving at a certain flow velocity in water have a 15 times greater Reynolds number than objects moving at the same velocity in air.

Nachtigall W. **Biomechanik – Grundlagen, Beispiele, Übungen** *Vieweg (2001)*

Completely different or comparable?

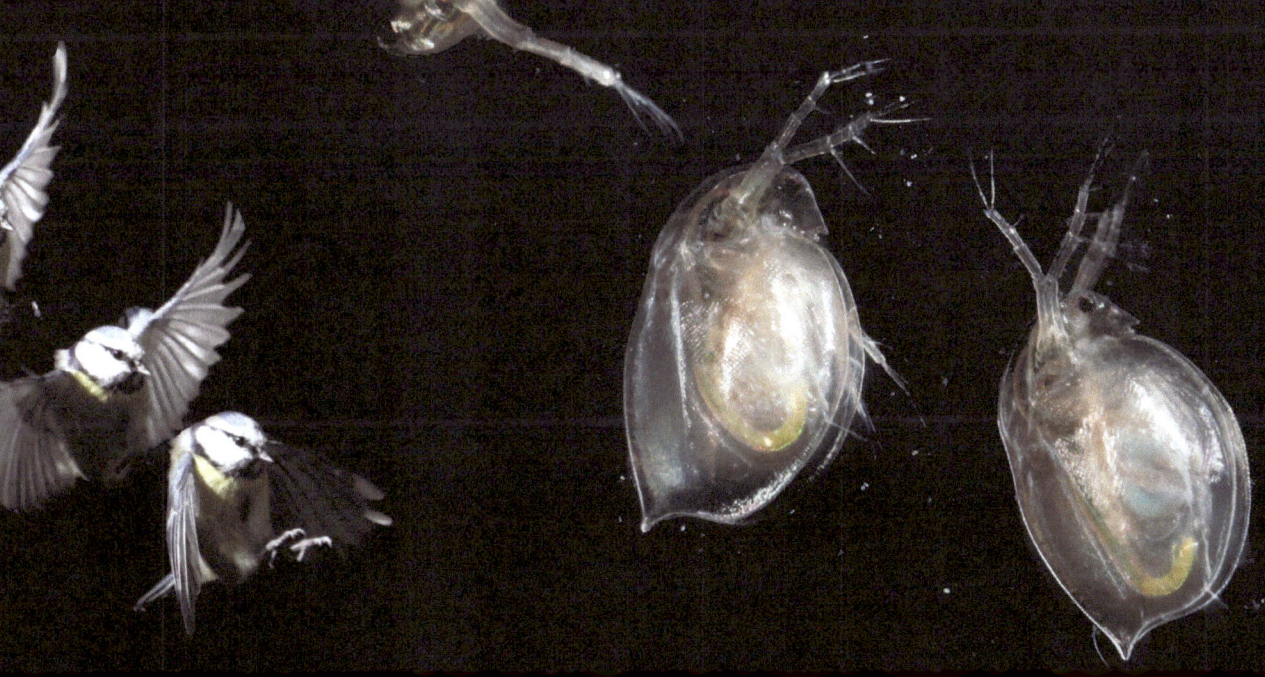

=> **Same** *Re*-**numbers** ⇒ **comparable conditions**

If two bodies have the same *Re* numbers, they are comparable in terms of fluid mechanics. This can be explained with the following example: Let's say a water flea – here, *Daphnia pulex* – moves at a Reynolds number of 300. Then its inertial force is 300 times greater than its viscous force. This figure is derived from a body length of $l = 3$mm, a "jump" velocity of $v = 10$ cm/s, and the corresponding value for the kinematic viscosity of water.

Flow ...

If one wishes to study a model of a water flea that is ten times larger, the surrounding water must flow at one-tenth of the velocity in order to arrive at the same *Re* number and the same balance of forces. However, such a low flow velocity in water is difficult to achieve to any degree of precision. Could a wind tunnel be a solution? Since the kinematic viscosity in air is 15 times greater than in water, the model must be surrounded by air moving at a flow velocity that is about 15 times higher, that is, at 1.5 m/s, which is easier to achieve. If the experiment succeeds, one could combine tests in water with tests carried out in air. One medium allows to determine one parameter, while the other medium lends itself to determining another parameter. Such procedures are commonly used in fluid-mechanical measurements. We will return to this point via a biological example during the discussion of the oil tank measurements for the study of the fruit fly's wingbeat (see page 62f.).

The cockpit

The head is multi-functional

The head, as shown by this close-up image of a gliding yellow-legged gull, may be regarded as the cockpit of the "bird aircraft". All sensors are bundled inside this head: the sometimes ultra-sharp eyes (eagles, vultures) – their sense of hearing, which may push the limits of the laws of physics (owls) – the flow sensors (covert feathers, small non-flight feathers, with several layers on the wings), refined olfactory sensors (noses), and sometimes even dynamic air speed indicators (petrels).

The head with its differently shaped beaks also acts as "food-intake device". The beaks of some seabirds are equipped with salt glands that excrete the salt in seawater. Moreover, the head is usually very flexible, which allows it to perform its many tasks.

The trunk "follows the head"

During flight as well, the head is not held rigidly but is often swung to the sides, probably as a way of scanning the terrain over which the animal is flying. This has been shown by footage from small cameras that were strapped onto tame eagles. Finally, the movements of the head also serve to signal the turning into a new flight direction by means of muscle spindle sensors attached to the head muscles of the flying animal.

Airflow along the wing: The basis for lift generation

Drawings of the main airflow of the wings in gliding flight show that there is a stagnation point somewhat below the bend of the wings. This stagnation point must be bypassed by the airflow above the wings. For this reason, the air must travel a longer path on the upper side of the wing than on the lower side – usually taken as the basic principle for the generation of lift (cf. the following double-page).

Groebbels, F. **Der Vogel als automatisch sich steuerndes Flugzeug** *Natur (38) (1930)*
Nachtigall, W. **Warum die Vögel fliegen** *Rasch und Röhring (1991)*

Yellow-legged gull (*Larus michahellis*)

A simplified representation

The air flows around the wings of gliding and sailing birds in the same manner as it does around the wings of an aircraft. The airflow splits at the stagnation point $St.$ The generation of lift is usually attributed to the asymmetrical shape of the wings. Since the upper surface of the wing is more strongly curved than the lower surface, the airflow at the top must travel a longer path and thus move faster at the top than at the bottom in order for the split airflows to meet again at the trailing edge of the wings.

Faster airflow ⇒ suction

As stated by Bernoulli, the faster airflow across the upper surface creates low pressure (suction). The wing thus generates lift by combining suction on the upper surface (approx. 2/3) with pressure on

Circulation and the Coanda effect

Aerodynamic engineers have pointed to the existence of circulation and the Coanda effect. Circulation refers to a counter vortex that rotates opposite in direction to that of the trailing vortex (see below), and the Coanda effect states that airflows tend to stay attached to the wing surface. This latter effect plays a role in ensuring that the airflow on the upper surface is sucked towards the airflow on the lower surface to produce the mass flux that is directed diagonally downwards.

1 The significance of the sharp trailing edge

Upon taking to flight, birds create velocity by flapping their wings. Airplanes must roll on the runway until they achieve a minimum velocity before take-off. The wings of airplanes, like those of birds, have an aerofoil profile, which is characterized by a sharp edge at its rear

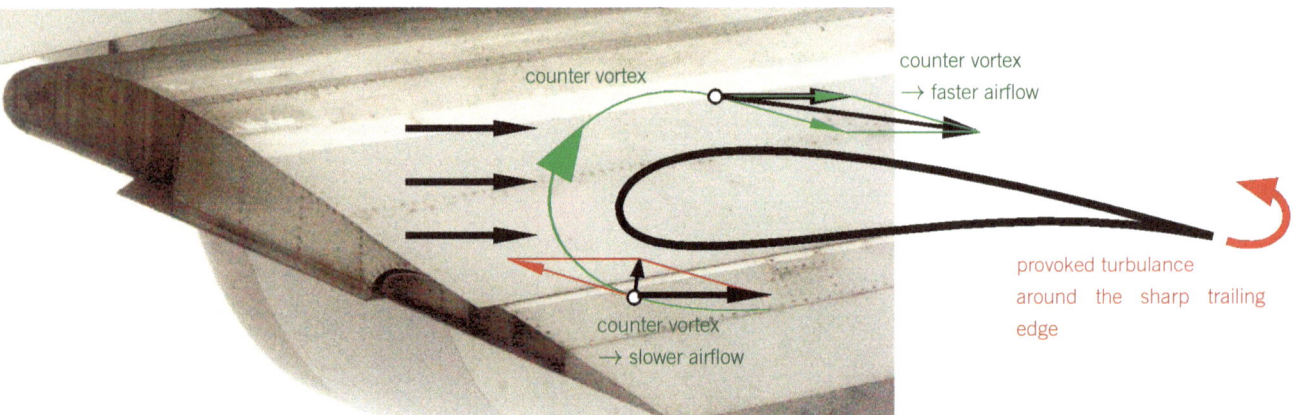

counter vortex

counter vortex → faster airflow

provoked turbulance around the sharp trailing edge

counter vortex → slower airflow

the lower surface (approx. 1/3). The following problem arises: The split airflows at the top and the bottom cannot meet at the trailing edge of the wings as long as lift is generated, as the airflow across the upper surface moves significantly faster. Therefore, this simplified version of Bernoulli's explanation cannot be valid.

Another theory goes back to Newton

Due to the downward curvature of the rear ends of the wings, the lower surface deflects and the upper surface sucks the flowing air, which has mass, towards the rear and downwards. The result is an opposite reaction of the air, which exerts an upward force on the wings. This vertical component of the airstream is known as lift.

Two different approaches to the same phenomenon

Both approaches produce the right results when using the actual parameters: that is, for Bernoulli's approach, the actual velocity vectors of the flows at the upper and lower surfaces (difficult to measure), rather than an estimation of the speed difference, and for Newton's approach, the actual measurements of the temporal mass flux.

Bill B.H.K. **Airfoil lifting force misconceptions widespread in textbooks**
    http://www.eskimo.com/~billb/wing/airfoil.html (15.2.2016)
Tennekes H. **The Simple Science of Flight – From Insects to Jumbo Jets**
    The MIT Press, Cambridge/MA, London (2009)
Anderson D.A., Eberhardt S. **Understanding Flight** McGraw Hill (2010)

end. Without this sharp trailing edge, it would not be possible for an airplane to fly.

2 Circulation

Circulation may be explained in the following manner: When an airplane takes off, a vortex forms at the trailing edge of its wings (drawn in the sketch rotating counter-clockwise). Following the law of momentum conservation, every vortex has its counter vortex. This vortex is shown in the sketch to rotate in the clockwise direction around the aerofoil. This counter vortex is referred to as circulation vortex, or simply circulation.

3 Manipulating the flow velocity

Due to its rotation direction, the circulation vortex decreases the flow velocity on the lower surface (velocity vectors shown in red) and increases the flow velocity on the upper surface (velocity vectors shown in green).

4 Pressure and suction effect

Bernoulli's principle can now be applied, creating pressure on the lower surface and suction on the upper surface, which then results in the generation of lift.

Suction is ultimately "pressure from the other side"

The term "suction effect" should be applied with caution. There can be no suction in a gaseous fluid like air, because "air columns" do not tolerate strain. What actually happens is that the positive pressure on the lower surface is increased by the negative pressure on the upper surface. Negative pressure always entails positive pressure from the other side.

Take-off and landing flaps

To increase lift during take-off, the wings of airplanes are equipped with slats on the leading edge, which act like a bird's alula (175), and with trailing-edge flaps, as shown in the image series. These flaps and slats increase the wing's curvature and thus enhance lift. This also plays an important role during landing because the velocity decreases and so does the lift generated by the airflow. Without such auxiliary devices, the airflow could break down and disrupt the generation of lift. Brake flaps can be extended on the wings' upper surface to reduce the aircraft's velocity during landing.

The oceanic whitetip shark (*Carcharhinus longimanus*) has a density greater than water. So, if the shark stops swimming, it will sink. During swimming, the stiff pectoral fins, which are slanted upwards, generate lift like an aircraft wing. Since the pectoral fins are positioned before the animal's centre of mass, the generation of lift results in a torque that tilts the shark's head upward.

However, the shark's dorsal fin simultaneously produces a torque that tilts the head downwards. Both torques compensate each other, and thus, it is possible for the shark to swim straight ahead without sinking.

Rigid wings …

Rigid wings, tucked away or not

Fruit flies of the genus *Drosophila* (left-hand page) carry rigid wings only a few millimetres in length, which are held in a roof-shaped position while they are at rest. The wings of rove beetles (Staphylinidae) are roughly the same length. They are also stiffened into a rigid position during flight (image series to the right). At rest, they are tucked away like cardboard boxes under the rove beetle's short elytra.

The flight of fruit flies and rove beetles is comparable

Due to the similarities in size, shape, and movement, it may be assumed that the wings of fruit flies and those of rove beetles function in the same or at least a very similar manner.

The flight behaviour of fruit flies has been extensively studied

Their wings have been recreated in the size of a tennis racket, and using a special mechanical system installed in an oil tank, these wing models have been made to swing as they do in their natural environment. The circular arrangement of images of a fruit fly on the left-hand page give a good idea of what a complete cycle of a fruit fly's wingbeat looks like.

"Reynoldian similarity" in the oil tank

Measurements in an oil tank have two major advantages: To begin with, it is possible to beat the larger model of a fruit fly's wings at a much lower speed – at 1 Hz, for instance, rather than the usual 300 Hz. If the settings of the oil tank model are adjusted properly, the physical characteristics of the flow should be the same in both cases (the similarity has been described and quantified by Reynolds). Secondly, the supporting structure of the wing model may be equipped with strain gauge strips, which can measure how the wings bend in all directions. This allows conclusions to be drawn about the magni-tude and direction of the forces acting in this process.

Different components in the generation of aerodynamic force

Measurements of this kind make it possible to determine the effect of stationary and non-stationary components in the generation of aerodynamic force. An aerodynamic force is stationary if it does not change over time. This is more or less the case with the downstroke of the wing, as there is little variation in the range and direction of the wing's movement.

A sudden turn in less than a millisecond ...

At the lower turning point, however, the wing rotates very quickly around its longitudinal axis. The angular velocities are astonishingly high, with up to 50.000 degrees per second (that is, 140 full rotations per second!), and the whole movement takes less than one thousandth of a second.

... provides one half of the necessary aerodynamic force

Such a process is highly non-stationary. It will accelerate the strong vortex diagonally downwards. This will produce a reaction force diagonally upwards. The "classic" stationary components provide sufficient force for large birds, but not for smaller insects. For fruit flies, the non-stationary components provide a staggering 50 percent of the necessary aerodynamic forces.

Rove beetle (*Staphylinidae*)

Dickinson M.H. **The effect of wing rotation on unsteady aerodynamic performance at low Reynolds** J. Exp. Biol., Vol. 192, 179-206 (1994)

When giants take to flight ...

A stag beetle taking to flight

This specimen was 7 cm long and weighed about 10 grams. Stag beetles (*Lucanus cervus*) are by far the heaviest European insects. So, when taking to flight, this arthropod struggles as much as a very large bird. The flight itself is relatively stable and straight, and it can reach a climb angle of 45°. Yet, this beetle cannot keep up with the agility of smaller flying insects.

The elytra flap as well

The wings beat 80 times per second in order to achieve the task of taking the stag beetle to flight. The images on this page show that stag beetles flap their elytra during flight. Elytra as generators of stationary aerodynamic forces had already been studied some time ago with the aid of precise aerodynamic two-component scales.

What is the maximum limit?

The size and weight of some tropical insects far exceeds even those of the heavy stag beetle. The distinctive feature of the longest beetle in the world, the Titan beetle with a body length of at least 20 cm, is already highlighted by its scientific name: *Titanus giganteus*. The biggest specimens are said to weigh up to 110 g, which raises doubts as to their capacity for *stable* flight.

Nachtigall W. **Zur Aerodynamik des Coleopterenflugs: Wirken die Elytren als Tragflügel?** *Verh. Dtsch. Zool. Ges. Kiel 1963, 319 - 326 (1964)*

One of the heaviest flying beetles

is the African *Goliathus cacius*, a flower chafer which may only reach a length of 10 cm, but can weigh up to 35 g (its larvae can even weigh 110 g). As can be observed with indigenous species of flower chafers (cf. pp. see page 40f.), these beetles do not open their elytra during flight. Instead, they spread their wings through lateral incisions on the elytra. See also p. 236.

The 3-cm long scarab beetle (*Scarabaeus*)

uses the same technique (several shots of the beetle taking to flight pictured above and below), which makes it difficult to tell when it is about to take to flight. It is remarkable that, right before take-off, the beetle will rotate in such a way that the sun shines on its back, as can be gathered from its shadow (S). This may actually be a navigation technique of the beetle (which uses the polarisation patterns of the sky for navigation when there is no sunshine).

Coordination …

Garden chafer vs. stag beetle?

The Garden chafer *Phyllopertha horticola*, which can often be found in June, and the stag beetle *Lucanus cervus*, which is only rarely seen in July, are both members of the scarab family. However, a large male stag beetle may weigh up to 100 times more than its smaller relatives.

Depending on the weather, both beetles like to fly – which may be less surprising for the smaller beetle, but it is still astonishing considering its "body shape". The stag beetle's capacity for flight is a great feat of evolution (cf. pp. see page 52f.), given that an animal's body weight increases to the cube of the magnification factor, while the wing surface – in the case of similar appearance – only increases to the square.

The take-off phase is critical

The moment of take-off requires aerodynamic forces great enough to keep the beetle in air and allow it to rise higher. So, lift must be greater than the beetle´s body weight. This is problematic because there is no existing airflow; the wings must generate this airflow by flapping. The beetle must beat its wings with as much amplitude and frequency as possible, as well as with the best possible linking of wing-beat and rotary oscillations. This, in turn, implies a great expenditure of energy.

Movements repeat themselves in the air

Once again, we draw a comparison between the two beetle relatives. Once they are in the air, the situation becomes easier – especially for the large beetle, which can fly at a relatively high speed (even for a person running on the ground, it is difficult to keep up with the beetle). Comparisons of high-speed photography series reveal striking similarities. Their motion sequences are rhythmical and periodical.

The upright flight position may be an advantage

The stag beetle must fly in a more "upright" position in order to maintain balance with its heavy pincers. This upright flight position has the advantage that the beetle can land on tree trunks with ease, which is what they are typically aiming at. The wing-stroke plane may be tilted slightly against the longitudinal axis of the beetle's body, but not by very much. If the beetle wants to fly slowly, it must increase the trunk's angle of attack, that is, raise its trunk to a steeper angle. The wing-stroke plane is thus tilted more horizontally, and as a consequence, the beetle generates more lift than thrust. In this manner, it is capable of keeping its body weight in the air and even rising higher, though this will lower its flight velocity.

Stag beetle (*Lucanus cervus*)

Velocity and body tilt

In general, the slower an insect flies, the steeper the angle of its body. This can actually be observed in fruit flies, which almost hover on the spot (cf. pp. see page 62f.), and in scoliid wasps of the genus *Scolia*, which fly slowly over dunghills in search of host larvae to deposit their eggs.

Different flight performances

Reluctance to fly

This jumping and flying great green bush-cricket (*Tettigonia viridissima*), one of the larger indigenous species of bush crickets, is a female, as can be seen by its long, slightly curved "ovipositor". It flies, in fact, only a few metres from one bush to the next. The slightly larger migratory locust (*Locusta migratoria*) from the family Acrididae can stay in the air for days and let itself be carried by the wind over several hundred kilometres.

Rayney R.C. (ed) **Insect flight** *Blackwell Sci. Publ., Oxford (1976)*

Jumping makes take-off easier

Sea birds sometimes take to flight by pushing themselves off with strong kicks and splashes of their legs. Insects often jump by using their hind legs. The important thing to bear in mind is that they must generate "airflow" at the very beginning of their flight. Jumping, in addition to the first strokes of the wings, contributes to this generation of airflow. It also helps to protect the tips of the wings from touching the ground or water.

Largest and smallest birds

Smallest bird

We do not know what the smallest fossilized birds may have looked like. The tiniest pipsqueak that can still be found whirring through the air is probably the bee hummingbird (*Mellisuga helenae*) with a wingspan of 9 cm and a body weight of approximately 2 g. With a wingbeat frequency of at least 50 Hz, this hummingbird can actively fly as long as its energy allows.

Largest bird ever known

The absolute giant among large birds may have had the capacity for flight, though only in gliding flight through Andean uplifts and thermal winds: *Argentavis magnificens*, which had a wingspan of 7.5 m and lived in Argentina some six million years ago, is assumed to be the largest bird to have ever existed. It is believed to have weighed only 72 kg and to have reached a velocity of up to 100 km/h in gliding flight.

Largest wingspan among living birds

One of the largest birds living today is the Andean condor (*Vultur gryphus*). On average, the Andean condor reaches a wingspan of 2.90 m, at a weight of 11.4 kg. There have been individual cases of male specimens that reached a wingspan of 3.10 m at a weight of 15kg. The Andean condor thus comes very close to the limits of the capacity for active flight. Its wingbeat frequency of 1 Hz may enable the large bird to take flight, but only for short durations. The condor depends on Andean uplifts to keep itself aloft for longer runs.

Flying with 7500 times as much weight, but at the same speed

The smallest and largest living birds differ by a factor of 34 in terms of wingspan but by a factor of 7500 when it comes to weight (extreme cases). It is interesting to note that the tiniest birds and the giants fly at approximately the same speed, around 50 km/h. What then are the upper and lower weight limits for flying birds?

Upper weight limit set by evolution

Birds, like all non-parasitic animals, obtain energy from their own metabolism. They produce a certain amount of energy per unit time – a process also known as metabolic performance. Flying requires a certain amount of physical effort, the so-called flight performance. Birds can fly in a particular flight condition as long as they are able to produce a metabolic performance greater than the flight performance necessary to realize that flight condition.

Both metabolic performance and flight performance increase exponentially with a bird's body weight. Flight performance starts at a lower level, but it increases more rapidly than metabolic performance. As it turns out, the two performance curves intersect at a weight of 12 kg. It takes a bird of 12 kg almost all of its metabolic performance to take flight while still covering the other vital functions of its body. The largest birds living today are the wandering albatross (*Diomedea exulans*) with a maximum size of 3.25 m and the previously mentioned Andean condor.

Evolution pushing the physical limits

It is interesting to see how close evolution has brought birds to the physical limits. The average body weight of the heaviest living birds that can fly (mute swan *Cygnus olor*, California condor *Gymnogyps californianus*, great white pelican *Pelecanos onocrotalus*, Kori bustard *Ardeotis kori*) lies at around 18-20kg.

These birds still have the capacity for active continuous flight, but birds with higher body weight, such as heavy male specimens of the bustard family, can only make short jumps into the air. Unless they are kept in the air by an external source of energy, which would ultimately be the sun, which creates thermal winds and other forms of uplift. Birds with a body weight of up to 15kg may be able to fly under such conditions, as illustrated by the Andean condor. The Andean condor is, in fact, an example of evolution that pushes the physical limits.

Amazilia hummingbird (*Amazilia amazilia*)

Campbell E.K., Tonni E.P. **Size and Locomotion in Teratorns (Aves: Teratornithidae)** *The Auk Washington DC100, 390-403 (1983)*
Nachtigall, W. **Vogelzug und Vogelflug** *Rasch und Röhring, Hamburg, Zürich (1987)*
Pennycuick, C. **Animal flight** *Studies in biology 33, London (1972)*

Evolution and physics

Griffon vulture (*Gyps fulvus*)

Lower weight limit

On the other hand, the smallest bird species do not have any problems achieving the necessary flight performance through their metabolism; they are "energetic giants," so to speak. Their muscles are more efficient per unit mass than the muscles of "morphological giants".

The smaller species may be "energetic giants", but …

However, a prerequisite is that they eat enough food to keep their metabolism running at a peak rate. Hummingbirds thus need to consume nectar around the clock. While they rest at night, they lower down their metabolism and body temperature in order to save energy (*torpor*). This is also the case with the smallest bird indigenous in our latitudes, the kinglet (genus *Regulus*). Despite their nocturnal torpor, they are barely left with energy in the morning, which is why they must immediately resume food consumption. Otherwise they would be dead in a matter of half a day.

Energetic relations, which ultimately mirror physical conditions – the relation between surfaces and body mass, for instance – thus govern an animal's life and its evolution in every detail. This also applies to bird migration.

"Ecological types"

It only makes sense to draw comparisons between similar "ecological types". Among flying birds, there is a maximum mass ratio of 7500:1. A common ostrich weighs up to 100 kg. Numerally speaking, the mass ratio between an ostrich and the smallest hummingbird species would be seven times greater. However, such a comparison makes little sense from a functional point of view.

Common ostrich (*Struthio camelus*)

Falcons: the fastest animals in the world

Gyrfalcon (*Falco rusticolus*)

A   B   C

Ponitz B., Schmitz A., Fischer D., Bleckmann G., Brücker C.
Diving-Flight-Aerodynamics of a Peregrine Falcon (*Falco peregrinus*)
PLoS ONE 9(2): e86506. doi: 10.1371/journal.pone.0086506 (2014)
Ciesieleski L.C. **Der Gerfalke** *Westarp Wissenschaften, Hohenwarsleben (2007)*

Gyrfalcon *(Falco rusticolus)*

Nosediving at 90 m/s and more

Right: The images on the right-hand side were taken at an interval of 1/60 of a second. The falcon in these images flies 2 m in 1/12 s against a static background, that is, at a speed of 24 m/s or 86 km/h – which is quite remarkable for a bird. The fastest bird is the peregrine falcon, which can reach a speed of more than 340 km/h (95 m/s) in a nosedive. In these images, the falcon flies after a lure that a falconer swings on a cord in circles before letting go. In the first images, the falcon is about to miss its turn and is struggling against centrifugal force, which is acting to force the bird out of its circular path, with its wings fully spread and held almost vertically upright. Towards the end of the image series, the falcon assumes its normal flight position as it drops its wings to a more horizontal angle and folds them slightly in. It is striking that the head is always held horizontally even when its wings are spread in an extreme vertical position. In the first images, the head is tilted away from the wings at a 90° angle. This allows the falcon to maintain spatial orientation even during such flight manoeuvres.

Detailed scientific studies

The most detailed studies (Ponitz et al 2014) have been carried out on trained peregrine falcons as they dive down a steep 60m high dam. They do not reach the same velocity extremes as in free airspace (up to 320 km/h), but they come close with a velocity of 80 km/h. Their peak acceleration was 1.2 g. During the first phase, the wings are almost tucked in; the transitional area from the proximal to the distal portion of the wing acts as a kind of secondary wing bend (A). The falcon navigates by slightly changing this posture. During the final phase, the legs and wings are tight against the bird's elongated body; the bird now forms a unique body that offers little resistance to airflow (B). Wind-tunnel models have been constructed based on such images in order to analyse their behaviour in high-speed tunnels. At the highest Reynolds numbers (cf. pp. see page 56f.), which correspond to a velocity of 144 km/h during nosedive, an astonishingly low minimum frontal area drag coefficient of only around 0.08 was measured.

Clumsy only outside of water

African or black-footed penguin (*Spheniscus demersus*)

From Antarctica to the equator

All penguins live in the southern hemisphere. The Humboldt pengu-in (*Spheniscus humboldti*), for instance, is endemic to the Pacific coast of the southernmost areas of South America. It is a member of the genus of banded penguins, to which the Galapagos penguin (right-hand page) also belongs. The Galapagos penguin is the only penguin species that can be found on the Galapagos Islands, and it may thus be said to live almost in the "northern hemisphere".

Incapable of flight but muscular

All penguins no longer have the capacity for flight, but their body shape and shortened wings are perfectly adapted to fast swimming underwater. Since penguins generate thrust on both the upstroke and the downstroke, they have equally strong upstroke and downstroke muscles. That is why their shoulder blades, to which the muscles that control the upstroke are attached, have a very large surface compared to the shoulder blades of other birds.

Galapagos penguin (*Spheniscus mendiculus*)

Heavy bones

In contrast to the hollow bones of flying birds, penguin bones are also solid and heavier because swimming does not require reduced weight. Penguins hunt for fish underwater, and for this purpose, they can dive for astonishingly long durations. They are mostly monogamous, and they hatch their two eggs on the bare frozen ground or inside a cave.

The transformation of covert feathers

The small, stiff covert feathers of penguins provide a scale-like, streamlined coating to their trunk and fins. Gentoo penguins (*Pygoscelis papua*) have thus been measured to have an incredibly low coefficient of front-face resistance of $c_{WSt} = 0.07$.

Thrust on the down- and upstroke

The photographs of Humboldt penguins pictured on the next double-page show how they flap their fins while swimming underwater. The fins have an almost symmetrical profile which allows them to be angled against the current, with the upper surface on the upstroke and with the lower surface on the downstroke. Due to its symmetrical profile, the penguin's fin generates thrust both on the upstroke and on the downstroke. There is no need to generate lift, as the penguin is perfectly balanced in water.

Lift and thrust during a hummingbird's flight

There are certain parallels with the flight of hummingbirds, which have also a symmetrical wing profile. During its hovering flight on the spot, the hummingbird generates lift both on the upstroke and on the downstroke, which balances their body weight. Thrust is not generated, as this would cause the hummingbird to drift away.

The bubble trick

In underwater footage in nature films, penguins that have just dived into the water can often be seen wrapped in a cloak of air bubbles. As they dive into the water, the layer of air that is attached to the surface of their body is stripped away. However, some air remains trapped beneath their short feathers. When the penguin returns from a long hunt and starts to sprint underwater in order to catapult itself off of a rock back into the air, it will also leave such a trail of bubbles. By pressing the covert feathers against the body, the air beneath the featherss is released. This acts like a form of boundary lubrication. For a moment, the penguin is surrounded by air instead of water, which is more viscous than air. This enables the penguin to reach the necessary velocity to leap out of the water. There is a type of Russian torpedo that releases air at its tips and can reach very high velocities in this same manner.

Nachtigall, W., Bilo, D. **Strömungsanpassung des Pinguins beim Schwimmen unter Wasser** *J. Comp.Physiol.B 137, 17 - 26 (1980)*

Fly?

Humboldt penguin (*Spheniscus humboldti*)

The small, stiff covert feathers of penguins provide a scale-like, streamlined coating to their trunk and fins.

Penguins "fly" underwater; these images show how they slap the water with the upper and lower surface of their fins.

Occasional fliers

They seem like gondola passengers, but ...

Spiders and mites do not have wings. However, there are some members of the class of arachnids that can actually travel through air. They spin a long thread that can easily be carried away by the wind and from which they hang like gondola passengers from a balloon. Of course, this comparison is not quite accurate, since these spiders do not fly using the principle of balloon flight. Balloons generate "aerostatic lift" because their gas filling is specifically lighter than air. Another force that keeps the balloon aloft are the vortices shed behind its body. This results in a pressure resistance $F_{W\,Pressure}$. The Reynolds number Re provides information (see page 56f.) on the relationship between pressure resistance and frictional resistance (cf. pp. 56f.). Yet, a spider's threads are heavier than air. So, how and on which principles do these threads manage to stay airborne in the atmosphere and even lift up their relatively heavy passenger?

times greater than the pressure resistance, which can thus be practically ignored. For the thread, the air is a viscous medium, which only causes friction. This friction, however, generates a high drag coefficient, which decelerates the thread's sinking velocity to a low value.

Frictional resistance predominates

In order to understand this, one only needs to look at the forces that act on the spider's thread. When the thread sinks in still air, it is hit by airflow from below. Every body that is enveloped by flow is surrounded by onion-like layers of that fluid that flow with varying velocities and thus causes friction to the body. This introduces a frictional resistance $F_{W\,Friction}$.

Shed vortices and high resistance coefficients

This is where the Reynolds number (cf. pp. see page 56f.), which is basically the quotient of pressure resistance to frictional resistance, comes into play;

$Re = F_{W\,Pressure}/F_{W\,Friction}$. For a thin thread with a diameter of 1/100 mm that sinks at a speed of 10 cm/s, the Reynolds number equals $Re \approx 1/15$: That is to say, the frictional resistance is 15

Every mild puff of air is enough for take-off!

If such a thread is caught by an uplift with a speed greater than 10 cm/s, it will be lifted up into the air. Wind speeds slower than this are practically never found. A slight breeze (wind force 2, "barely noticeable on the face") already reaches a speed ranging from 1.6 to 3.3 m/s. That is, every wisp of wind no matter how soft it is will carry the spider's thread and its weight in whatever direction the wind is blowing. Spiders thus travel very comfortably and without wasting much of their own energy, using their threads in accordance with the laws of physics.

This is how adult spiders of the family Linyphiidae typically migrate in autumn, and their young ones do so in early summer. Some mites (spider mites, Tetranychidae), which also belong to the class of arachnids, also travel in this manner.

As regards the image: Once this little cross spider lets go and releases its thread, it will automatically be lifted up into the air. Yet, this behaviour is quite unusual for cross spiders.

Roberts, MJ. **Spiders of Britain and Northern Europe** Collins Field Guide (2001)
Foelix, R.F. **Biology of Spiders. 3. ed.** Oxford Univ. Press (2011)

Gliding without wings

Even some snakes ...

The five species of flying snakes, which reach up to 1.2 m in length, can flare out their ribs like the flying dragon *Draco volans* (cf. p. 11). They thus extend the surface of their body to twice its size, and the lower side of the body is curved like an aerofoil. Upon take-off, they assume an S-shaped body posture and slither through the air, which aids in generating lift.

...and ants and spiders can glide!

The *Cephalotes clypeatus* ant is equipped with a shield on its head, which it may use for gliding when it jumps from a tree. *Odontomachus bauri*, a Panamanian ant that can swim at high speeds, uses its hind legs as stabilisers, while its front legs and middle legs are alternately used for rowing. Only recently has it been discovered that some species of spiders (genus *Selenops*) can glide in a similar manner.

Socha, J.J. **Kinematics - Gliding flight in the paradise tree snake** *Nature 418 (6898): 603–604 (2002)*
Socha J.J., O'Dempsey T., LaBarbera M. **A 3-D kinematic analysis of gliding in a flying snake, *Chrysopelea paradisi*** *J Exp Biol. 208(Pt 10):1817-1833 (2005)*
Yanoviak S.P., Dudley R., Kaspari M. **Directed aerial descent in canopy ants** *Nature 433 (702): 6624–6 (2005)*
Yanoviak S.P., Dudley R. **The role of visual cues in directed aerial descent of *Cephalotes atratus* workers (Hymenoptera: Formicidae)** *J. Exp. Biol 209 (Pt 9):1777-83 (2006)*
Yanoviak S.P., Munk Y., Dudley R. **Arachnid aloft: directed aerial descent in neotropical canopy spiders** *Journal of the Royal Society Interface 12 (110) (2015)*

Detailed analysis of a landing flight

A little bit of physics can be very insightful

Especially with smaller birds, the landing flight can happen in a flash, and it is difficult to make out any details with the naked eye. For this reason, such a landing flight shall be examined in close detail here, using biophysical methods of slow-motion analysis.

85 frames per second

The image series was shot at 250 frames per second. Pictured here and highlighted in the graph on the next page is every third frame, that is, the frames number 0 – 3 – 6 etc. Die interval Δt that has elapsed between one trio of frames and the next thus equals about 1/85 s.

Velocity and acceleration diagram

Since it is not possible to determine a centroid, the centres of the eyes in subsequent images have been drawn over each other to create a distance-time graph $s(t)$. Between each trio of frames, the average landing speed $v = \Delta s/\Delta t$ has been calculated by difference analysis and are plotted as $v(t)$ both in metres per second and kilometres per hour. An approximate delay $-b = \Delta v/\Delta t(t)$ may also be calculated and plotted as $-b(t)$ in metres per square second and in units of earth acceleration g.

Braking wing flaps to reduce velocity

This comparison is quite instructive (The numbers used in the evaluation are rounded to two decimal places, as is common in technology; the accuracy of the evaluation on the basis of the drawings is, of course, considerably lower). The blue tit lands in a slightly wavy line, which is inclined downwards at an average angle of no more than 15°. Shortly before frame 0, the bird makes a braking flap with its wing, then two more on its way to the landing site, and a third weak flap upon touching down with its feet (frame 18, cf. drawing). One would think that the landing velocity is abruptly reduced with each braking flap of the wings. As the graphs show, this is the case to some extend. The braking flaps cause fluctuations in the $v(t)$-curve, but the fluctuations are not particularly strong. These flaps may produce some kinks in the curve, but in general, the landing velocity decreases at a constant pace. This is due to the inert mass of the bird, which puts a damper on any rapid changes.

The velocity is reduced to half within 1/14 of a second due to the braking flaps

Between frames 0 and 3, the blue tit is landing at a velocity of about 2.34 m/s or 8.42 km/h. Up to the moment when the bird touches down with its feet, that is, over a flight distance of 12.70 cm, which the tit crosses in 0.072 s, aerodynamic braking alone causes the velocity to be reduced to 1.21 m/s or 4.36 km/h, that is, to its half.

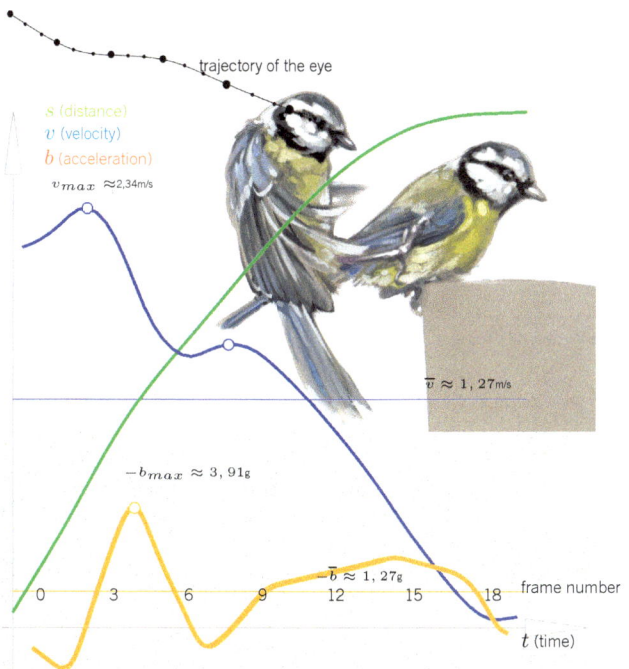

Above: computer evaluation of the image series at an interval of 1/250 s. *Green: distance-time-diagram, blue: velocity-time diagram, orange: acceleration-time-diagram (mirrored on the abscissa). Every third frame (1/83 s) is numbered. Initially, only these frames were "manually" evaluated, but the final outcome already became apparent at this point.*

Now the shock absorbers come into play

In the remaining images until the bird comes to rest upon landing (frame 18 – final position), the work is done by the bird's legs, which are initially stretched out at an angle of 145° and are gradually folded in until they hang at an angle of about 45°. These legs are astonishingly effective as landing devices and shock absorbers.

The delay can only approximately be determined by means of "manual" differential calculations. Due to the braking flaps of the wings, it fluctuates considerably. Over a flight distance of 12.7cm, that is, until the moment of the feet's touch-down, the tit slows down from 2.34 m/s to 1.21 m/s within 0.072 s. The average delay thus equals $-b = (2.34 - 1.21)/0.072 = 15.69$ ms^{-2} or 1.60 g.

Up to 4g delay in the air

When comparing individual frames, however, the delay may vary considerably, namely between 22.10 ms^2 (which equals 2.25 g) within the first trio of frames and 41.65 ms^2 (which equals 4.25 g) within the final trio. So, over this short distance of braking flight, the delay changes by a factor of 1.89. During its landing, the bird even accelerates slightly, but this may not be the case in other pictures of a blue tit's landing flight.

Now the bird must touch down!

Upon touching down with its legs, the bird slows down further until it comes to rest. The deceleration forces thereby generated must cushion the impact of the landing legs. It has already been mentioned that the legs are bent in by about 100° upon touch-down. The leg extension muscles, which are already activated and thus prepared for the strain, as well as their tendons, are further stretched in the process, like previously stretched tension spring, on which further weight is attached. At the same time, the bird tilts clockwise around its lateral axis and thrusts its head forward until it comes to rest in the position drawn with dashed lines. In the final phase, other muscles, especially those in the back, also help to alleviate the impact of the touch-down. The energy absorbed in the process is released as heat into the environment unless it is used to change the position of the bird's trunk. The   in the final phase amounts to an average of 2g. Good "shock absorbers" are needed to alleviate such an impact!

Perfect interplay of sensory organs and muscles

The seemingly effortless landing process does not just "happen by itself," but it must be precisely controlled and navigated, which is enabled by a complex interplay of sensory organs acting as sensors on the one hand and flight, leg, and trunk muscles acting as effectors on the other hand. And this must all take place within a timeframe of just a few milliseconds.

Hunting for insects in the air

Insects must be caught during flight

The picture-perfect European bee-eater hunts for bees, dragonflies and other insects in flight. This requires fast reactions, which also come in handy during mating season, when the males compete with each other in flight battles. Fast changes of direction and altitude are part of their repertoire.

Rising speed of 1.8 m/s

The next page shows images of an upward flight at an angle of almost 60°, which was shot at 240 frames per second. When viewed as a diagonal, one can discern 12 positions at an interval of 1/20 s each. The eleven intervals between the individual positions have been increased by 10. This has been done to avoid overlaps of the images, because the bird only crosses about 22 cm in one tenth of a second, which corresponds to an average rising speed of 1.8 m/s. The images

show a full wingbeat cycle, that is, a downstroke and the subsequent upstroke.

"Animal-fixed and space-fixed trajectory"

In the sketch above, the measured points of the wingtip relative to the animal's trunk are connected by a line that represents the so-called "animal-fixed trajectory". The sketch illustrates the direction of the wingbeat from the rear top to the front bottom and then back up on the other path (the upstroke path lies behind the downstroke path). At the lower turning point, the trajectory makes a loop. This loop is due to the movement of the flexible primary feathers, which whip back again as the distal part of the wing is raised. At the upper turning point, both paths meet at almost the exact same point. One wingbeat thus functions like the next. The corresponding "space-fixed trajectory" indicates the subsequent positions of the wingtip in space,

Bee-eater (*Merops apiaster*)

as the European bee-eater ascends diagonally from the first image to the last, and looks similar. It extends more into space than the animal-fixed trajectory, and gives an impression of how the wings function like a propeller during a perfectly vertical rising flight.

Wings functioning like a horizontal propeller

A propeller draws air from above and accelerates it downwards. This create an upward reaction force, which lifts the bird. In the present case, the "propeller" stands diagonal within space, and therefore, creates an additional reaction force that is directed to the front, thus generating thrust as well as lift. When transitioning to horizontal flight, the space-fixed wing layer is held at a more slanted angle to produce more thrust. This enables the bird to travel at its terminal velocity.

Like a vertical take-off aircraft on a carrier

Vertical take-off aircraft with variable-pitch propellers (VTOL-system, suitable for aircraft carriers) work according to the same principle; first they generate only lift and then they will additionally generate more and more thrust.

Analysis of a rising flight

Animal-fixed tra-
jectory (red) and
space-fixed trajecto-
ry (black)

Details of a European bee-eater's (*Merops apiaster*) rising flight: Full cycle of a wing-beat, intervals increased by ten towards flight direction. The bird crosses about 22 cm within this short time frame (1/11 of a second). Analysis on the left-hand page.

White-fronted bee-eater (*Merops bullockoides*)
(an African bee-eater species)

Bats and birds as a role model

Seychelles fruit bat (*Pteropus seychellensis*)

A mixture of a bat's hand and the primary feathers of large birds

Leonardo da Vinci (1452 – 1519) famously designed flying machines, which could be strapped onto a person's back in order to fly like a bird. The wings of these devices look like the wings of bats or flying foxes. They consist of bent rods, which resemble a bat's hand. Impregnated canvas is stretched between these rods, not unlike the flying membranes of a bat. Yet, bats were not the only role model in the design of Leonardo's flying machine. The cunning bird-watcher adopted a principle of surface formation that can be observed with the primary feathers of large birds. The barbs of these feathers are placed into an eccentric position relative to the shaft. As a result, the primary feathers of large birds form a closed surface on the downstroke, whereas, on the upstroke, they open up again and let the air flow through. Leonardo da Vinci imitates this principle by using a system of overlapping lids for his wings.

Although it does not comply with the laws of statics and fluid mechanics ... These designs may have been inspired by nature "on paper", but they could not function because they did not comply with the basic laws of statics and fluid mechanics. Back then, these scientific disciplines only existed as empirical observations. Moreover, Leonardo underestimated the problem of driving power, as in many of his inventions, such as a self-propelled cart that anticipated the modern automobile. In order to generate sufficient lift with such wings, a person would need at least 25 times as much mass of pectoral and arm muscles. Of course, this has done nothing to diminish the fascination inspired by the inventions of this ingenious observer and interpreter of nature.

Seychellenflughund (*Pteropus seychellensis*)

The profile of a bat's wing differs from that of a bird

The wings of bats and flying foxes do not have the same "profile" that characterizes a typical bird's wing and that is decisive in effectively generating lift (cf. pp. see page 56f.). They still impress with their excellent flight performance, especially during long-distance flights (cf. pp. see page 52f.).

This may probably be attributed to the leading edges of their wings. The leading edges are not curved like the leading-edge radii of bird wings, but sharply keeled due to their membrane's stiffening. As a result, the boundary layer of air is rendered turbulent and cannot detach itself so easily from the wing's upper side (which would lead to a collapse of lift). In addition, this creates a leading-edge vortex, which has a positive impact on the flow at the upper side. Special kinematics also allow bats to use their wings when they are inclined at high angles. This may explain their extremely good manoeuvrability, which is, in turn, the reason behind their extraordinary predatory behaviour.

Evolution optimizes living systems in their entirety

So here as well one complements the other. Evolution always optimizes living systems in their entirety, though it does so by "rigorously exploiting" the physical possibilities and limits. Figuratively speaking, evolution is not about optimizing the boundary layer or the like, but rather about bringing together "optimizations of all imaginable aspects" of an organism in way that renders the organism as a whole as viable, competitive, and reproductive as possible, or in the Darwinian sense, in a way that renders the organism as "fit" as possible.

4 Criteria of evolution

Sexual selection
Climatic changes

One reason why evolution works so well is the existence of two sexes. These two sexes are joined in a variety of ways. In most cases, the female choses a male based on certain fitness criteria, which the male must demonstrate (female choice). The male must often directly compete for the female through battle (male-male competition). If the next generation is supposed to fly better, then a major role is also played by mutations that permit the possibilities of biophysics to be stretched even closer to their limits.

Sexual evolution (1)

Sexual selection is particularly effective

While natural selection encompasses all aspects of everyday life, such as finding food, protecting oneself against enemies and the daily competition for resources of any kind, the competition in finding and obtaining a sexual partner is subject to a particularly tough process of selection known as sexual selection. In general, sexual selection can be defined as follows: It contributes to or causes the development of those behavioural patterns and structures that increase an individual's chances for successful mating (copulation).

Higher chance for mating

The sole concern of sexual selection is how I can achieve a higher frequency and chance for mating than my species mates. It comes into effect once the individuals of a species are mature and ready to copulate. So, put very simply, one could say that natural selection is about survival in order to make it to the final battle for selection, namely the battle of obtaining a sexual partner successfully.

Sexual selection is very immediate

Since this is not so much about survival but about ensuring the perpetuation of one's genome through the offspring, sexual selection is very immediate. The most obvious manifestation of the effects of sexual selection is a sexual dimorphism that causes a species' males and females to look distinctly different.

Two types in sexual dimorphism

A distinction is made here between two basic types:

1. Intrasexual selection:

Selection between partners of the same sex

a) male-male competition,

b) female-female competition

2. Intersexual selection:

Selection between partners of different sexes

a) female choice,

b) male choice

Different forms of sexual dimorphism

These two basic types of sexual selection produce very different forms of sexual dimorphism. Male-male competition induces the development of all kinds of fighting structures. In vertebrates, this is often reflected by the development of horns and antlers. In insects as well, the development of various types of horns can be observed on the head and/or on the thorax, as is the case with the European rhinoceros beetle and the stag beetle with its mandibles that resemble antlers. What they all have in common is the fact that these horns and antlers are used to compete for females. Red deer stags are particularly well known for battling for the right to mate with as many females as possible. A similar behaviour can also be observed among Galliformes, such as the chicken. Cocks fight for hens using a spur on their hind feet.

Nuptial plumage of various kinds

Female choice, on the other hand, favours the development of nuptial plumage among males. To our human eyes, the nuptial plumage of male peacocks and birds of paradise might look grossly extravagant. The exact look of such nuptial plumage depends on the individual species, where they live, which sensory organs they preferably use, and other aspects. That is why there is even acoustic and olfactory plumage. Their common function is to demonstrate a male's fitness as a parent. For this purpose, these feathered adornments, as well as the way they are presented, sometimes accompanied by scent and singing, must be lavish and elaborate. The idea is that only fit males are capable of developing such extravagant, and sometimes quite obstructive, structures.

Nuptial plumage is thus an indicator of fitness

which causes the female, in a competition involving several males, to pick the one that appears to her to be the fittest for mating. So what underlies the development of such nuptial plumage is fierce competition and selection. The fact that such attributes are primarily found among males rather than females is due to the fact that females are considered by the males to be a scarce resource for which they have to compete. **Not all females are ready to mate**

Although the gender ratio of most species is usually balanced so that there is a female for each male, only a small fraction of these females are actually available to the males. Females may be either not in estrus (not ready to mate) or pregnant.

Common ostrich (*Struthio camelus*)

Stag beetle (*Lucanus cervus*)

Sexual dimorphism among birds

Flight and gliding organs may also exhibit several forms of sexual dimorphism. The wings of male birds, as well as other parts of their bodies, are often decorated with very elaborate ornaments. These ornaments sometimes consist of specially modified feathers on the wings (great argus), their backs (peacocks) or their tails (chicken). These ornaments, however, may push the limits of what is convenient and manageable, as is the case with the peacock. The male peacock must grow a huge fan of feathers with ever more perfectly developed eye-shaped spots in order to impress the females, but it must still be able to escape a tiger with this long train of feathers.

The first dinosaur feathers may have been courtship ornaments

In debates on the origins of bird wings, there has been the hypothesis that the first feathers may have been used as ornaments for courtship displays in addition to having a protective function. These feathers attached to the front limbs of dinosaurs may have gradually become larger and more colourful, which, in turn, led to a lengthening of the front limbs until these feathered front limbs eventually turned into wings that could first be used for gliding and ultimately also for flying. The flaw in this hypothesis lies in its failure to justify why it is not only males that developed the capacity for flight.

African moon moth (*Argema mimosae*)

Therefore, only few females are actually available at a given time, and there is fierce competition for these females. To improve its chances in this competition, the sex that has to compete for access to the other must make a bigger effort to attract a mating partner. These are usually the males, because, as mentioned, there are usually fewer females ready to mate than males.

Male insects often hatch first

Among many insects, which typically have a short lifespan, the competition is enhanced by the fact that the males hatch long before the females. Once the first females hatch, the males of a colony are already up and eager to get hold of these females for themselves. This increases the competitive pressure enormously, and the process of sexual selection thus also involves efforts by males to be the first to find and copulate with such females.

Selection for sensory organs

This implies a selection for sensory organs. Animals that look for mating partners using their visual organs typically have particularly big compound eyes; animals using their olfactory senses undergo a process of selection that affects their antennae, which are the source of this sense. That is why the males of such species are equipped with more complex antennae than their female species mates. A well-known example are the male antennae of many moths (e.g. the African moon moth), which can smell female pheromones from a distance of several kilometres.

Sexual evolution (2)

Why does the pin-tailed whydah have such long tail feathers?

Or the peacock? Why does the red deer develop such an enormous pair of antlers year after year? Growing and moving such antlers and feathers require metabolic energy, which is ultimately a valuable asset. Such ornamental features do not bring any noticeable benefits to the everyday lives of these animals. So, why has evolution produced and preserved such seemingly wasteful structures?

Pin-tailed whydah (*Vidua macroura*)

Long-tailed paradise whydah (*Vidua paradisaea*)

Darwin, Charles **The Descent of Man, and Selection in Relation to Sex (1st ed.)** London, John Murray (1871)
Anderson, M. **Sexual Selection** Princeton Univ. Press (1964)

The question of "why?" often arises,
But there is no scientific way of answering it. Problems that can be approached from biology or physics, can only raise questions that ask "to what extent" and "in what manner". To what extent are the long tail feathers beneficial to the pin-tailed whydah (*Vidua macroura*)? We can make observations about it. We do not ask why something that didn't exist before developed, but we, instead, start with the fact

that something is there and consider what it might be good for. Such questions allow us to advance hypotheses that can be further explored. The question of "in what manner" thereby serves as a heuristic principle.

The males of the pin-tailed whydah have been observed to use their long tail-feathers as objects of attraction during courtship dances, which they perform in front of a seemingly uninvolved female. Further comparisons and observations show that, when it comes to selecting a copulation partner, females prefer to pick those males that have the most sumptuous plumage: This can also be seen in experiments in

which the tail feathers are artificially lengthened, which makes these males more attractive to females.

It is ultimately only the reproductive success that counts
If evolution is assumed to lead to the "survival of the fittest" – as Darwin's principle is often misinterpreted – then the male that manages to reproduce and pass on its genes wins the competitive struggle against the other males as "the fittest." This male qualifies as the fittest even if its courtship ornaments require much metabolic energy and are rather inconvenient for their daily lives, and even if the muscles of said male are comparatively weak so that it must frequently hide from its rivals. What ultimately counts in evolution is only the reproductive success. All developments and modifications crafted by evolution point towards this common aim.

Mating of flying animals

From top to bottom: storks, common kestrels, doves, sparrows
Right-hand page: mallards, flying foxes

"Intermediate stages of evolution"

All instances of mating may be regarded as intermediate stages of evolution, since they pave the way for new genetic combinations.

Diploid and haploid chromosome sets

The animals pictured on this pages are all diploid, possessing a double set of chromosomes. Following cell division, however, their egg and sperm cells are haploid, carrying only a single set of chromosomes. Copulation, as rough as it may appear sometimes, must guarantee a safe passage of sperm. A sperm cell can then merge with an egg cell in order to create a diploid chromosome set.

The difference to the somatic cells of females:

The fertilized egg cell now carries the genetic information of mother and father in equal parts. The young one is thus "genetically different" from both its mother and its father.

The males fight for the females

The competition among species mates to achieve higher reproductive success is particularly tough when it comes to obtaining a mating partner. This battle is usually carried out in two ways: The males often fight face-to-face for the possession of a female. Birds like the New World blackbirds pictured here fight using their whole body, their beaks, and their feet. Other animal species have developed fighting structures that they can use as weapons to fight against other males for predominance. Male stag beetles have enormous branched jaws that look like antlers, and they use these jaws to turn an opponent on its back and drive it away from a female. The winner may then copulate with the female.

Like their ancestors, domestic horses live in larger or smaller herds led by a stallion. The stallion must fight for and defend his alpha position. With horses that are kept outdoors, such as the Icelandic horse, males and females are typically separated to avoid such fights between male opponents. During breeding season, however, the studs return to the herds of mares, which may trigger fierce battles. The biological function of these battles is to ensure that the mares mate only or primarily with powerful and healthy studs in order to increase their foals' chances of survival.

Male hares box and beat their rivals fiercely in order to keep them from having access to the waiting females. Since the females give birth 3 to 4 times a year, from February until well into autumn, they are ready to mate once they finish suckling their young. For this reason, the males must invest considerable effort in order to become fathers as many times as possible.

Interceptors

During the mating season of ducks, drakes can often be seen flying after each other and playing tricks on their opponents. This can less frequently be observed in geese – pictured here a Greylag goose, *Anser anser* – but if they engage in such a behaviour it is no less dramatic. In this picture, goose no. 1 flies after goose no. 2 – both geese are probably gander. Goose no. 1 manages to push goose no. 2 out of the way and overtake it in flight, forcing goose no. 2 to the ground. Goose no. 2 appears to be smaller in size and weaker. This Greylag goose may have been tired from the beginning, as it gives up relatively soon. In the third pair of images, it already draws its wings into gliding position, and in the fourth pair, the goose veers off and starts to glide to the ground. The "air battle" is thus over. Goose no. 1 is likely to fly a loop now and return to its point of departure, without following its opponent any further, as it has already achieved its goal. Any further pursuit would only be a waste of energy.

A crash-landing from exhaustion

The fleeing and exhausted goose probably wanted to land as soon as possible. So, it glided at a steep angle and high velocity. Due to the remaining momentum, the goose tilts slightly forward upon touching down with its feet. In order to avoid falling on its nose, it pulls its wings to the front and supports itself on the area of its elbow joints and on its primary feathers that are pressed flat to the ground.

Sexual dimorphism

The Victoria crowned pigeon (*Goura victoria*) from New Guinea is a ground-dwelling pigeon with a feather crown on its head. Both sexes have such a crest and so this cannot qualify as an example of sexual dimorphism. The noticeable diversity between the two sexes is a particularly visible expression of the effects of sexual selection, and it is all the more pronounced if the males have to compete fiercely for the females.

The bond between two animals is usually limited to mating and sometimes also to the joint raising of the offspring. Once this has been achieved, the two sexes separate. In species where life-long bonds are formed and monogamy predominates, both sexes typically look the same. The Victoria crowned pigeon is a monogamous pigeon and is, therefore, not marked by sexual dimorphism.

This does not apply to the Indian peafowl (*Pavo cristatus*), which is a markedly polygamous species of galliforms. That is why their sexual dimorphism is particularly pronounced. The males must go the extra mile, so to speak, in order to be selected by a female. For this purpose, sexual selection has promoted the development of their downright excessive plumage, which may, hence, be seen as a consequence of the tough competition for females.

Since all males compete for females and since the females are prone to mate with those males that have the longest tail feathers and thus more space for their distinctive eye-shaped spots, sexual selection will favour the development of this feature until natural selection sets a limit. Long colourful feathers may be beneficial in attracting females, but they are rather counter-productive when it comes to escaping from enemies.

Vertical take-off bird

The images on this page illustrate the positions of the individual primary feathers during the flap of the wings. A peacock must rise high in order to flaunt its marvellous plumage …

A feather grid for more lift

In such instances of high lift generation, one can easily imagine how every single feather is enveloped by airflow like a small wing. Together, these feathers form a kind of grid that offers additional fluid-mechanical advantages.

Elongated upper tail coverts

Adult males have a tail that consists of about 150 feathers with an eye-shaped spot at the tip of each.

A classic example of sexual selection

The example of the peacock illustrates very clearly that sexual selection may have favoured the evolution of exorbitant plumage to impress females, but natural selection strikes with no mercy once the peacock finds itself in a situation in which it must escape quickly, which is made considerably harder by its train of long feathers. This will ultimately limit the impact of sexual selection.

Shimmering colours

The shimmering feathers contain no colour pigments. The impression of colour results from light refraction caused by tiny air-pockets within the barbs of their feathers.

A duet of hovering flight

Ruddy darter (*Sympetrum sanguineum*)

The males fly at the front …
A male dragonfly clasps a female of the same species by her neck with its anal appendages, and he then continues to fly with her in tandem. Copulation has thus taken place, and the female then looks for a suitable spot to lay her eggs.

…and drag the females until exhaustion.
To lay their eggs, the females of some species dip the tip of their abdomen into the water, others land on and inject each individual egg into leaf material, and still others submerge themselves underwater to lay their eggs on aquatic plants. These latter regularly drag the male, which does not release the female, along with them into the water.

Willow emerald damselfly (*Chalcolestes viridis*)

"Male guarding" to avoid sperm competition
Copulation has already been achieved. The male still clasps onto the female while she lays her eggs, because rival males may try to grab the female by its head and remove the sperm from the previous mating (see also p. 104).

Brown hawker (*Aeshna grandis*)

Piggybacking parasites

Pretty parasites

These little parasitic flies (here *Gymnosoma rotundatum*) from the family of tachina flies (Tachnidae) are no longer than 88 mm. Upon copulation, the females lay their eggs on the skins of shield bugs. The larvae then bore into the host's body and hibernate there. In spring, the full-grown larva breaks out of the host, pupates, and hatches later as an adult fly. The host insect survives in most cases.

Avoiding sperm competition

If a female immediately mates with another male, the previous male runs the risk of having his sperm removed by the second male, which will replace it with his own sperm. So, in the process of sexual selection, some animals exhibit behavioural patterns that are intended to avoid sperm competition (see also p. 102).

The pair of flies can still fly with the male piggybacking on the female. Immediately after copulation, the male remains mounted on the female long enough for his sperm to fertilize the eggs that the female is about to lay.

Robert L. Smith (Editor) **Sperm Competition and the Evolution of Animal Mating Systems** *Academic Press (2012)*
Simmons, L.W. **Sperm Competition and Its Evolutionary Consequences in the Insects** *Princton Univ. Press (2001)*

Dangerous spines to defend their territory

The European wool carder bee *Anthidium manicatum* is a solitary bee species. The males are armed with five spines situated at the tip of their abdomen, which are used to defend their mating territory against food competitors. Their territory typically contains many flowering plants favoured by this bee species. The males patrol and defend these patches of flowers in hovering flight, and they chase away food competitors, such as honey bees, bumblebees and other males of the same species (see image above). They curve their abdomen forward as they approach their enemies so that their spines are directed forward. The aggressive males may, however, injure their fragile wings in the process.

The favourite plants are "reserved" for the females

The females, which gather pollen and nectar in their territory, are typically approached by a male to copulate on their favourite flowers (bottom right). The most successful males are the ones that attract the most females with this strategy.

Lampert, K.P. et al. **'Late' male sperm precedence in polyandrous wool-carder bees and the evolution of male resource defence in Hymenoptera** *Animal Behaviour* 90: 211-217 (2014)

Signals from both sexes

Male: a bright colour signal

In ad¬dition to their bright yellow colouration, the males of the Cleopatra butterfly (*Gonepteryx cleopatra*) have – in contrast to the common brimstone (*Gonepteryx rhamni*) living north of the Alps – a large orange patch on the upper side of their forewings. This patch also reflects ultraviolet light and thus acts as a flashy colour signal to attract females.

Female: "Snubbing" with their body language

The males perform a courtship flight to present their patches in front of a sitting female.

If the female is willing to mate, it will lower its abdomen. The abdomen of the female that is pictured here sitting on the blossoms of a chaste tree (*Vitex agnuscastus*) is clearly raised, signalling a rejection of the male.

The butterflies hibernate

Similar to the common brimstone endemic to our regions, these butterflies living in the South hibernate and then reappear in early spring in order to lay their eggs on buckthorn species. The caterpillars feed on buckthorn leaves. The new generation then emerges in June.

Sex, drugs, and butterflies

Within the group of predominantly tropical monarch butterflies (Danaidae), an interesting courtship behaviour can be observed. In order to impress their females, the males must gather toxic substances that ooze from injured plants. These toxic substances are usually pyrrolizidine alkaloids, from which the males can produce the pheromone dihydropyrrolizidin. During courtship, this pheromone is transferred onto the antennae of the females by means of a brush-shaped, eversible organ at the tip of their abdomen known as coremata. The image of the paper kite (*Idea leuconoe*) on this page clearly shows the two everted coremata. Copulation can only occur if the male produces these pheromones, in as great a quantity as possible.

Once again sexual selection ...

This is another example of how sexual selection can cause extravagant behavioural patterns among males, whereby the females can indirectly measure the male's potential as a breeding father. The possession of dihydropyrrolizidin pheromones is an indicator of their fitness.

Boppré, M. **Sex, drugs, and butterflies** *Natural History 113: 26-33 (1994)*
Boppré, M. **Leaf-scratching – a specialized behaviour of danaine butterflies for gathering secondary plant substances** *Oecologia (Berl) 59: 414-416 (1983)*

Ophrys – the plant that has sex

An *Anthophora balearica* bee (Mallorca) has been deceived by *Ophrys dyris*. The pollinia are attached to the tip of the bee's abdomen.

Species-specific fragrance mixtures

During the mating of most insects, the females attract their males across long distances by means of species-specific fragrance mixtures, the sexual pheromones. These mixtures are specific to every species of insects so that the males only respond to pheromones of their own species, which prevents mating between different species.

Which male reaches the female first?

Throughout their lifespan, male bees spend the whole day looking for females that smell of those pheromones and are thus ready to copulate. Once a male smells a female of its own species, it will follow the trail of scent, and as it gets closer, it will locate the female by means of visual cues. However, other males will have discovered the same female, and now it depends on which male reaches the female first. This results in a pronounced sexual selection affecting the sensory organs involved in that race to mate, as males with better olfactory and visual organs are more likely to come out as winners and achieve copulation.

Orchids: Imitation of pheromones and mimicry

Some orchid species, such as the members of the Mediterranean genus *Ophrys*, have evolved in a way that allows them to twist this mating system to their own use. By imitating the sex pheromones of female insects and mimicking their visual cues, they attract male insects looking for a receptive female.

The trick works and leads to "pseudocopulation"

Male bees are actually deceived by this mimicry. They follow the trail of scent produced by an *ophrys* flower, land on the orchid's labellum, and try to copulate with the pseudofemale. This kind of behaviour is also known as pseudocopulation.

Contact with the pollen is inevitable

The orchid's petals are perfectly equipped for the visit of the male, because, with their size, colours, shape, and even with their simulation of a female bee's hair, they deceive the male long enough to get its body – either its head or the tip of its abdomen – into contact with its pollen.

Paulus H.F., Gack C. **Pollinators as prepollinating isolation factors: Evolution and speciation in Ophrys (Orchidaceae)** *Israel J.Botany 39: 43-79 (1990)*
Paulus H.F. **Wie Insekten-Männchen von Orchideenblüten getäuscht werden – Bestäubungstricks und Evolution in der mediterranen Ragwurzgattung Ophrys**
  in: Evolution - Phänomen Leben (759 pp.), Denisia (Linz) 20: 255-294 (2009) http://www.landesmuseum.at/pdf_frei_remote/DENISIA_0020_0255-0294.pdf
Ayasse M., Schiestl F.P., Paulus H.F., Ibarra F., Francke W. **Pollinator attraction in a sexually deceptive orchid by means of unconventional chemicals**
  Proc.Roy.Soc.London B 270: 517-522 (2003)

All of the pollen sticks to the insect

For this type of pollen to transfer from one flower to the next, orchids have developed a particularly sophisticated design. All of their pollen is concentrated into so-called pollinia. Upon contact with an insect, the whole pollinia will get stuck to its body (see image on left-hand page).

The olfactory signal is "not trivial"

Detailed studies of the *ophrys'* fragrance in comparison with the sexual pheromones of the simulated insect has shown that the orchid perfectly imitates the insect's pheromones. This imitation is not trivial, because pheromones consist of a mixture of 12 to 20 components. The visual imitation of some *ophrys* species is no less stunning, and is reflected by names such as bee orchid or bumblebee orchid.

Each *ophrys* species has its own pollinator

Such studies have also shown that each *ophrys* species has its own pollinator which distinguishes it from other species of the same genus. This can be attributed to their imitation of an insect's sexual pheromones, which, as previously discussed, are also species-specific.

The extracted pollinia stick to the body

of the deceived male (see left-hand page), and when the male is once again enticed into pseudocopulation, they are transferred onto another *ophrys* flower for pollination and fertilization.

Imitating a bee's patterns of fur growth

Another trick of *ophrys* orchids consists in, not only attracting male bees, but also causing them to either enter the plant head first or approach it backwards for copulation. For this purpose, these *orchid* species grow fine hairs on their labellum that run from front to back or vice versa: They thus imitate the growth pattern of a bee's body hair.

The males rotate on the flower

The males immediately detect the growth direction of the flower's fine hairs and rotate accordingly, as they would do with a female bee after picking up the tactile cues of her hair, which serve as an indicator of where the front part of her body is and where her back is. As a result, *ophrys* species which attract bees that engage in copulation from a reverse position typically stick their pollinia to the tip of the insect's abdomen, while the others deposit pollinia on their pollinators' heads (see image below).

Male of the Scollid wasp *Dasyscolia ciliata* (Scoliidae) pseudocopulating with the blossom of an *Ophrys speculum* (in Mallorca).

Sand bee *Andrena variabilis* copulating on *Ophrys kedra* (Crete) in reverse.

Spectacular aerial acrobatics

Air battles ...

Air battles are quite common among birds of prey. They usually do not go beyond feint or brief attacks, but they still involve spectacular aerial manoeuvres, such as inverted flight, sharp bends, and "side-slipping" into a steep spiral dive.

... and courtship flights

It is very probable that the two common buzzards in this "multiple image" are mating (males and females only differ in size, but not in appearance). During such a courtship flight, the two partners typically hook their claws. If the birds do not separate on time, though they might slam to the ground, as happened to the buzzard pictured on this page.

Such courtship rituals can also be observed in American bald eagles.

Common buzzard (*Buteo buteo*)

Long-legged buzzard (*Buteo rufinus*)

"Acceleration machine"

Like pheasant ...

The pheasant is a typical "acceleration machine," as Konrad Lorenz once labelled it so poignantly. That is: The pheasant cannot fly for long durations and at high velocities, but it can accelerate extraordinarily fast. Shortly after take-off, the pheasant reaches its "terminal velocity," and is thus able to escape many a dangerous situation. The fans of its spread primary feathers are particularly noteworthy, as is the extreme twisting of its wings in the middle image, which helps generating thrust.

... like cock.

The cock generally acts in the same way, though such manoeuvres usually serve the purpose of establishing his "supremacy" over his rivals rather than enabling a quick escape. The cock pictured to the left flutters its wing to rise up to a three-metre-high barn roof.

Natural aggressive drive

The right-hand page shows scenes from a (harmless) cockfight that illustrate the cocks' aggressive drive. If two cocks are made to fight against each other, it usually ends with the death of one of the two cocks. First written evidence of a cockfight dates back to China 517 BC, and such ritualized fights may have led to the domestication of chickens. While forbidden in almost all countries around the globe today, cockfights are still considered a popular "sport" on the Philippines. Cocks use a sharp spur on their metatarsal bone. The metatarsal spur of older cocks is very long and is used with much force. If the inferior cock cannot escape, as is the case in cockfights held in a ring, such fights can reach a deadly conclusion.

Aggressive behaviour

Attacking mute swans

(Cygnus olor) are not to be trifled with. They can also be dangerous to swimmers, as they might sit on top of them and push them underwater. During courtship or when it senses that its young are threatened, it takes off at a roaring start with powerful flaps of its wings. These flaps, which produce large splashes on the water, serve to increase their take-off velocity. In addition, they allow them to drive away small "enemies" that actually present no danger, such as mallards. These ducks will usually just fly up and land again in a spot nearby shortly afterward. The lower image shows how a drake places its wing like a parachute in brake position. The mute swan is one of the heaviest birds with the capacity for flight, as can be gathered from its ponderous take-off.

The swan's powerful take-off flap

(right-hand page below) touches the surface of the water and thus changes the current. The stagnation point on the lower surface must be located very far to the back, as the lower wing layers are splayed out to the bottom front, which is a rare sight. This tells us that the airflow must have started on the lower surface and must then have travelled along the bend of the wing to the upper surface. The pressure that is generated at this moment on the lower surface probably constitutes a major part of the total pressure.

Feeding and flying lessons

Below: A falcon mother has caught a vole, brings it back to the nest, and flies away again. A week later (see image to the left on the right-hand page): The young have grown and can barely be distinguished from their mother. Now the mother must teach them how to fly and hunt for prey.

Image to the right on the right-hand page: Falcons like to sit on telephone and electrical wires. This position allows them to locate their prey from above, and they can easily take off for their hunt. Sometimes they just "let themselves fall," in order to pick up speed and generate aerodynamic force.

In the image to the bottom right, however, the falcon actively takes to flight. The image illustrates how the bird's "surface" multiplies by opening its wings.

Global warming (1)

Praying mantis (*Mantis religiosa*)

Active spreading or recent introduction

The up to 8 cm long praying mantis (*Mantis religiosa*) is actually a Mediterranean animal. In Central Europe, this thermophilic species was previously only found in Southwest Germany (Kaiserstuhl region) or in the very east of Austria. In the last 10 to 15 years, however, the species has spread considerably. From the east of Austria, the praying mantis has travelled far to the west, along the valley of the Danube, and can even be found in the river's side valleys. In Germany, there are now stable populations in many areas of Baden-Württemberg, Rhineland-Palatinate, and Saarland. For several years, *Mantis religiosa* have been found in Bavaria, Saxony, Saxony-Anhalt, and even in Berlin. This species lives primarily on dry slopes, even on those that emerged only recently. It is difficult to assess if this regional expansion is due to active spreading or recent introduction. Yet the fact that this species can survive in these new regions and form stable populations may be ascribed to global warming.

The females do not eat the males!

In late summer, females lay brown cases of eggs in protected places and die shortly thereafter. After hibernation, the nymphs start hatching in April and May, and by shedding their skin throughout the summer, they grow to become adult mantis. As ambush predators, they sit waiting in grass and catch insects once they get within the reach of their raptor claws. Mating also takes place during summer. Contrary to popular belief, the male is not usually eaten by the female after mating.

Bees do not hibernate

The images on the right-hand page – honey bee (*Apis mellifera*) and small tortoiseshell (*Aglais urticae*) – were taken at the end of December. The pollen "trousers" of the bees have barely developed and are shiny – the flower must have produced a lot of nectar, but only few pollen. Bees do not hibernate: When the first frost arrives, they form a "winter cluster" around the queen. The bees at the outermost layer generate heat through vibration of their muscles. The food that the bees have gathered in summer and autumn and stored in the combs provides sufficient energy for this effort. In this way, the bees manage to keep the inside of the cluster at least at room temperature.

When the outside temperature rises above 12°C,

the bees make a cleansing flight to empty their intestines and then return to form a cluster once again. Butterflies like the small tortoiseshell and the European peacock, which often hibernate in protected places, can also be seen on such warm days, giving us the first glimpses of spring.

Once again "the warmest month on record" …
The photographs on this page where shot on a terrace in Vienna in December 2015. Fauna and flora immediately react to rising temperatures, but they may suffer setbacks due to cold spells in spring.

Small tortoiseshell (*Aglais urticae*)

Honey bee (*Apis mellifera*)

Global warming (2)

The impact of global warming on fauna

For several decades, a global climate change has been taking place, having serious repercussions for humans, plants, and animals. One classic example: The polar bear, which moves on surfaces of frozen water in order to catch its prey, is losing its natural habitat due to the melting of sea ice in the Arctic. A rise in temperature by only two degrees will dramatically accelerate the melting of Greenland's ice sheet.

Malaria mosquitoes

The malaria parasite is a protozoa (*Plasmodium*) that lives in in the midgut of *Anopheles* mosquitoes, which pass the disease onto humans as they suck their blood. However, the parasite requires high temperatures to complete its cycle. There are currently no infected mosquitoes in Europe, but this could change with the impact of global warming.

The mosquito pictured below is a species of the genus *Aedes*. The species of this genus, which are common in tropical zones, are vectors of many tropical diseases such as dengue fever, yellow fever, Japanese encephalitis, chikungunya, West Nile fever, etc. Two species of these mosquitoes have recently been introduced to Central Europe: Asian tiger mosquito (*Aedes albopictus*) and Asian bush mosquito (*Aedes japonicus*). Initially, these species were

Introduced via international transport,

which may, in turn, be related to climate change. However, the rising temperatures accelerate the development of these dangerous protozoa and increase the risk of spread.

Ticks and tick-borne diseases

Ticks are spreading further north and to higher altitudes due to global warming, and they remain active during warm winters, which accelerates their life cycle and leads to a higher population density. In recent years, a significant increase in infections with the tick-born encephalitis and Lyme disease has been registered, which must also be attributed to the increased survival rate of small forest rodents that are the natural hosts of these pathogens.

Asian mosquito (*Aedes*)

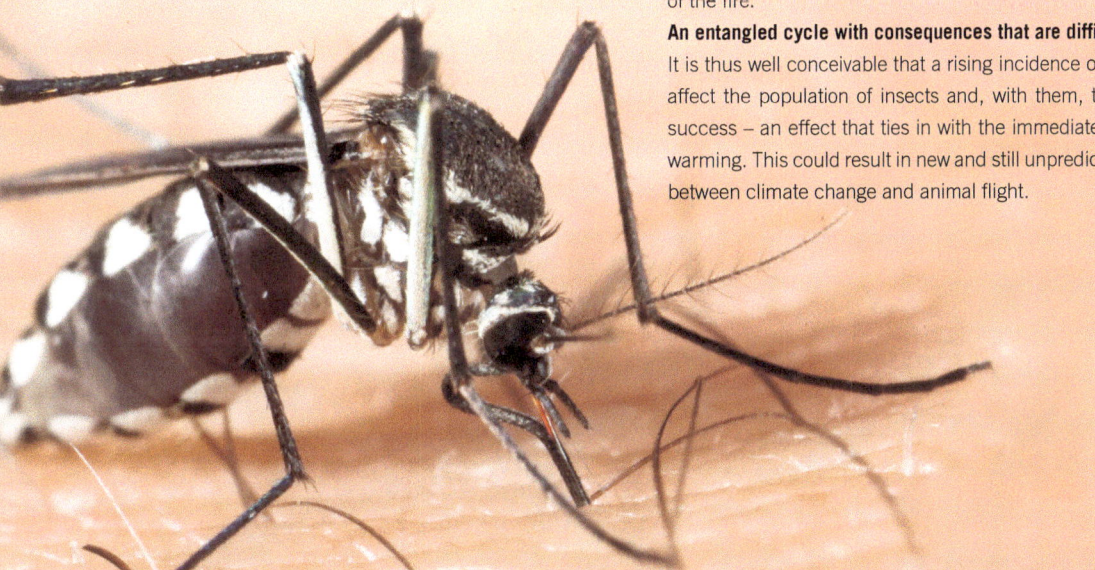

Dry conditions cause forest and steppe fires

Global warming has increased summer dryness in southern Europe, leading, in turn, to a dramatic rise in the incidence of forest fires. This increased dryness has also become noticeable during the summer months in Central Europe, where there have even been forest fires in early winter. The summers of Central Europe are predicted to become even hotter and drier in the future.

Flying away! But even that can be dangerous ...

Only animals with the capacity for flight have a chance to escape the deadly danger of forest fires. While animals with little mobility, such as insects in the larval stage are usually among the first victims. Only those who can fly away fast or burry themselves deep into the ground can survive such a fire. Yet even those that escape by flight may fall prey to predators, especially birds that gather near the front of the fire.

An entangled cycle with consequences that are difficult to predict

It is thus well conceivable that a rising incidence of steppe fires can affect the population of insects and, with them, their reproductive success – an effect that ties in with the immediate impact of global warming. This could result in new and still unpredictable interactions between climate change and animal flight.

5 Insects: The first flying animals

Using every niche

Insects are the most successful class of invertebrates. Most of them have a more or less well-developed capacity for flight. This capacity proved to be such a major advantage in their evolution that it evolved relatively quickly and consistently across the class. Even giant insects like the African Goliath beetle and the stag beetle can fly.

The evolution of insect flight

Firebrat (*Thermobia domestica*)

Insects as a model for success – but why?

Among the invertebrates, insects may be regarded as the model for success par excellence if measured against their enormous number of species. Their diversity is estimated at over 10 million species around the world. These are huge numbers when compared to mammals with about 5.600 species or birds with nearly 10,000-12,000 species. Thus the question arises as to what it is that makes insects so successful. On the one hand, insects are phylogenetically a very old group, which have had a lot of time to evolve since the Silurian period 443-419 mya. On the other hand, they have managed to spread across all kinds of environments, the only exception being the sea. This widespread distribution of insects was possible because their emergence is closely linked with the conquest of land by animals: The presence of the first plants created a wealth of new possibilities for life to thrive and evolve.

Insects for well over 400 million years

The earliest fossil remains of very primitive insects date back to the Lower Devonian almost 410 mya. They belong to a species of the so-called Apterygota from the still-living order of springtails (Collembola), and this species is known under the scientific name *Rhyniognatha hirsti*. Since these fossil remains already exhibit much resemblance to modern springtails, the actual emergence of primitive insects can be assumed to have occurred much earlier.

Apterygote insects

Today the so-called apterygote insects comprise the previously mentioned springtails (Collembola); the minute coneheads (Protura), which live deep in the soil; and the two-pronged bristletails (Diplura), which include species that can reach a body length of up to 5 cm and are, like earwigs, equipped with pincers at the tip of the abdomen. All three orders are distinguished by having their mouthparts almost fully enclosed within their head so that only a tip is visible. They are, therefore, grouped together as Entognatha. Further orders of apterygotes are the jumping bristletails (Archaeognatha) and the silverfishs (Zygentoma).

Tails to throw themselves up

The members of both orders have three long tails at the tip of their abdomen, which, on the one hand, act as antennae at the back and, on the other hand, allow these animals to throw themselves up by pushing themselves off the ground with their abdomen and then land smoothly. This flight mechanism can often be observed in the jumping bristletail (Archaeognatha). Silverfish live in cracks and crevices in the ground, under tree bark, beneath stones, and even in our flats and houses. They are of particular significance in the present context, as they have served as a model to illustrate a common theory on the development of insect wings.

Springtails (*Collembola*)

First hinge joints on the road to success

The second subclass of insects comprises the highly derived group of winged insects (Pterygota). Phylogenetically, silverfish are actually more closely related to the Pterygota than to the jumping bristletails, with which they share some resemblance. This classification has been justified on the grounds that the silverfish have already evolved double-hinged mandibles. That is, their mandibles are attached to the head capsules by two joints called condyles, which function as a hinge joint. Their ancestors, on the other hand, possess mandibles with only one condyle acting as a ball-and-socket joint. Hence, a distinction is made between Monocondylia and Dicondylia, with the latter including silverfish and all Pterygota. The functional advantage of these dicondylic mandibles is that they facilitate feeding on hard materials. The evolution of such structures has certainly contributed to the great success of insects.

More than 99 % of insects can fly!

The group of still-living apterygotes may comprise around 10.000 species, but these numbers are small compared to the vast amount of modern insect species that do not belong to this group. The lasting success of insects may actually be attributed to the development of wings, with which they were able to conquer the air. This large group of insects are known as winged insects or Pterygota. Apterygota comprise all those insect species that are primitively wingless.

A dire need for fossil remains

As with other flying animals such as birds and bats, the question arises as to the phylogenetic origins of insect wings and insects' capacity for flight. The lack of transitional forms among recent species poses some difficulties when trying to answer this question. That is why one must turn to fossil remains in order to shed light on the origins of insect flight. The present system of insects can also be studied in a way that allows us to draw conclusions about the origins and development of such structures.

All flying insects have a common ancestor!

Numerous recent morphological and molecular-genetic studies have provided some insight in this respect. An international team of 98 scientists has recently published a well acclaimed study on the phylogeny of insects, in which whole genomes were examined as to their phylogenetic significance. Relevant for our present context is the discovery that all Pterygota or flying insects are derived from one common ancestor. This implies that insect wings have evolved only once, and this must have happened during the Carboniferous some 300-350 million years ago.

"All of a sudden," giant dragonflies with a wingspan of 70 cm

Fossil remains from the early stages of this evolution at the end of the Silurian / beginning of the Carboniferous are astonishingly rare. There is much evidence, however, dating back to the Carboniferous period in general. We know, for instance, of the existence of giant dragonflies with a wingspan of up to 72 cm during this period. *Meganeuropsis permiana* (72 cm) from the early Permian in North America and *Meganeura monyi* (70 cm) from the Carboniferous in France are the largest fossil remains of insects found to date. The enormous body size of these insects is attributed to the elevated oxygen levels during that period. Like their present-day counterparts, these dragonflies were probably also voracious aerial predators. It must therefore be assumed that there was already a rich variety of flying insects serving as prey during that time.

Misof B. et al. **Phylogenomics resolves the timing and pattern of insect evolution** *Science* 346 (6210) 763-767, DOI: 10.1126/science.1257570 (2014)

Animal size ...

Primitive vs. more highly or most highly developed

We will compare a primitive order of insects (dragonflies, Odonata) with a more highly developed one (straight-winged insects, Orthoptera) and a most highly developed one (two-winged insects, Diptera).

A comparison yielding interesting results

● As the higher development of the animal increases, its body size and the length of its wing decreases,

● the differences between forewings and hindwings become more pronounced,

● the wingbeat frequency increases,

● wing propulsion transitions from being direct to indirect

● rather than having a unified muscular system, their muscles are divided into two separate systems responsible for wing propulsion and flight navigation respectively.

A) Dragonflies – a primitive order of pterygote insects

Dragonflies, which can still be quite large in their modern-day form, have forewings and hindwings that are similar in size and shape. Each of these four individual wings are controlled by direct flight muscles that produce the raising and lowering movements. Signals from the central nervous system will activate the forewings for a downstroke and the hindwings for an upstroke. The wingbeat frequency ranges around 20 Hz. This direct control of their wings enables dragonflies to be very agile during flight despite their very primitive neuromechanics, which had already evolved in their giant ancestors during the Carboniferous period.

B) Grasshoppers – a more highly developed order of insects

Grasshoppers are a relatively small species within the group of straight-winged insects and possess stiff forewings that can only be folded longitudinally and fan-like hindwings. In addition to direct flight muscles, they also possess indirect muscles that control the opposite wing and rotate it into different positions and locations. Small specimens can reach a wingbeat frequency of around 40 Hz. Due to the fine-tuned differences in wing form and flapping movement produced by their muscle system, these animals can also be fairly agile.

C) Two-winged insect – a most highly developed order of insects

Among the Diptera, that is, two-winged flies and mosquitoes, only the forewings act as aerofoils – the hindwings have evolved into halteres that no longer have any aerodynamic function and are only used as measurement organs for flight stability. The largest specimens achieve a wingspan of only a few centimetres. The wingspan of smaller specimens is no more than a few millimetres. Their wingbeat frequency ranges between 50 Hz for the large crane fly Tipulidae, 150-250 Hz for medium-sized flies and bees, and over 1000 Hz for the tiny fungus gnats Mycetophilidae.

Indirect wing propulsion necessary

Such high wingbeat frequencies, which are required for small specimens for physical reasons, cannot be achieved with direct muscle systems controlling wing propulsion. Evolution has thus lead to the development of an indirect system. The strong flight muscles attached to the thorax are of a fibrillar type. They are coordinated in such a way that the thorax is expanded and contracted at each motion of the wing and can thus oscillate – probably in resonance – at a high frequency. The wings attached to the sides are indirectly moved in the process: To the left and to the right, they are pulled up and down at the same rhythm.

Basic vibration according to the "wooden spoon principle" …

One can imagine this process by thinking of a pot covered by a small lid that slightly flaps up and down. If two wooden spoons are clamped laterally between the edge of the pot and the edge of the lid, leverage will cause the spoons to be drawn along by the movement and to swing up and down as well. This is how the basic vibration is generated.

… fine control by means of a system of very fine muscles

Fine control of the wings is achieved by means of a special system of very fine muscles, most of which are attached to the base of the wings. The engine and navigation system are thus completely separated: One is responsible for powerful flight, the other for an insect's exceptional agility.

Nerves "fire" only every four to five contractions

The oscillation of the flight muscles is faster than the rate at which the nerves can transmit signals. Such insects are said to have neurogenic or asynchronous control of their muscles, as the nerves fire only every four to five contractions in order to keep the muscles vibrating (as opposed to myogenic or synchronous control, where nerve impulses and muscle contractions are synchronized). Physical necessities and evolutionary developments of form and function thus interact closely.

Transport of goods

Pieces of meat are carried in front with the mandibles, sometimes with the assistance of the front legs. To keep itself from tipping over, this insect of the Hymenoptera order must produce a tilting moment that raises its head by means of specially adapted wing movements.

Common wasp (*Vespula vulgaris*)

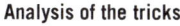

Analysis of the tricks

We will take a closer look at the take-off flight of a wasp that is about to carry away a piece of meat (image series to the left):

Splaying out the hindwings

At the beginning, the wasp is still in contact with the ground. The image series shows an upstroke, the subsequent downstroke, and another upstroke, which were shot at 500 frames per second. The shadows in image 1 illustrate that the hindwings are slightly spread away from the forewings. Air passing between this gap runs across the upper side of the smaller hindwings, avoiding a collapse of airflow. These wings may be compared to the flaps mounted at the trailing edge of an aircraft's wing, and they help generate the lift necessary for take-off.

"Near clap and fling"

A second modification that contributes to the generation of lift can be seen in image 3, and it may remind us of the "clap and fling" motion of butterflies (see page 56f.). The piece of meat must weigh more than half as much as the wasp itself, coming close to the maximum weight that the insect can carry in flight. So the transport of such goods requires high-lift generation.

Wasps are protective animals. The approaching wasp is also after the meat, and spreads its legs both ready to land and bite with its mandibles. However, once it detects the nest odour of its species mate, the wasp reverts to a more peaceful state.

Spider wasp (*Auplopus carbonarius*)

Right at the limit

Carrying if possible ...

Wasps – pictured here, a black spider wasp with a jumping spider it has just captured – usually carry the spiders they paralyzed to their nesting holes by moving backwards, walking undeterred by any obstacles. This can take an exhausting amount of time, and may require the wasp to make detours and find new approaches to reach its destination again and again. Without a doubt, the spider weighs considerably more than its predator.

... but also flying for short terms.

The photograph at the bottom left shows that the wasp may also use its wings while carrying its prey; in this case, as a "lifting aid" as the wasp walks up a vertical wall with the spider in its mandibles. The wasp can even carry its prey during flight, as shown by the photograph at the top left, which was taken shortly after the wasp's take-off. The wasp flies with its body upright at a steep angle and its wings flapping in a horizontal plane, and the mass centre of the prey, which the wasp holds against its abdomen, lies below the central point of application of the lifting forces. Hence, the wasp may not have any difficulties maintaining stability while carrying its load, but it may face problems when it comes to power. The transport requires all the power that the wasp can muster, and only barely so. This can be gathered from the short duration of these transport flights, which are usually no more than short jumps into the air. In between its jumps, the wasp must go back to the ground and rest.

One can even hear it

The sound of the wasp during flight is noticeably raspy, which points to the fact that the airflow is already partially interrupted on the upper surface of the wings. So the wasp's transport flight is right at the limit, something that evolution only barely managed to bring about.

Right-hand page:

Transporting food to the burrow

A digger wasp has brought a bush cricket under control by paralyzing it with its sting. To take its prey to the burrow it dug beforehand, the wasp must fly a distance of half a metre. The wasp seizes its prey, in which it will lay its eggs, "by the collar," and drags it into the burrow, where it will eventually serve as food supply for the wasp's larvae.

Digger wasp (*Sphex haemorrhoidalis*)

The most popular insect

Famous in many respects

Honey bees, *Apis mellifera*, are social insects with complex behavioural patterns. This is not confined to how bees communicate with their famous "dance language" and how they navigate between their hive and food sources. Their behaviour during the gathering of food is also astonishingly peculiar. Bees carry nectar and pollen to the hive to feed their larvae.

Transport of heavy loads

A honey bee can transport a pollen load of up to 35 % its own weight, which equals about 15-20mg. In addition, there is the nectar that they store in their honey stomach. When leaving the hive, the bee must calculate the amount of fuel it must consume in the form of honey in order to carry this load and get back to the hive.

When bees return to the hive after collecting pollen, they calculate how much energy they have used up on their way to the flower.

Calorie metre with stopwatch

Bees can measure their potential energy consumption with astonishing precision. Not only are their calculations based on measurements of the distance, but their brains are also equipped with a kind of calorie metre combined with a stopwatch. Experiments have shown that bees calculate the flight distance that they have tra-

Western honey bee (*Apis mellifera*)

velled and their efficiency during the search for food separately, and they can communicate both results independently from each other to their hive mates via different means, one of them being their dance language.

Crailsheim K. **Intestinal transport of sugars in the honeybee (*Apis mellifera* L.)** *J. Insect physiol.34, 839-845 (1988)*
Shafir S., Barron AB. **Optic flow informs distance but not profitability for honeybees** *Proc. Royal Soc. B Biol. scci 277 (1685), 1241-1245 (2009)*
Kremer F. **Zur Steuerung der Abflugmagenfüllung bei der Honigbiene (*Apis mellifera*)** *Zoo. Jb. Physiol. 85, 249-265 (1981)*
Harano K., Sasaki M. **Adjustment of honey load by honeybee pollen foragers departing from the hive: the effect of pollen load size** *Insectes Sociaux, 62(4), 497-505 (2015)*

How does the pollen get to the larvae?

Nectar is sucked through a tube-like tongue before being stored in the bee's honey stomach. Pollen, on the other hand, is brushed from the stamens onto the bee's body and is then passed from the forelegs to the midlegs and thence to the hindlegs. The bee then combs out the nectar and gathers the pollen in baskets at the tibia of its hindlegs. For this purpose, the insect uses a pollen brush, that is, fine rows of bristles on the inner surface of the leg's first joint, also known as tarsus.

The complicated process of "filling the pollen baskets"

The process of combing out resembles human hand-washing, with the fingers of one hand interlocking with the fingers of the other hand. This can be observed from behind once the bee leaves a flower. The so-called "spur" on the tibia brushes against the "spines" on the basitarsus, pushing one load of pollen after another from the opposite leg into the basket, until the latter is filled to the brim. Before being placed into the basket, the pollen is mixed with nectar from the honey stomach in order to give it a slightly sticky consistency. The fully loaded bee looks as if it is wearing wide trousers. These pollen trousers can reach at least one fifth of the bee's body mass, and so the flight back to the hive requires a higher metabolic performance of around 40 mW/g.

Nachtigall W., Feller P., Jungmann R., Rothe U. **Measurements of the metabolic power of the honeybee (*Apis mellifera* L.) during tethered flight** *Biona report 6, 81-87, Akad.Wiss. Lit. Lainz (1988)*

The honeybee phenomenon

(1) Drone
(2) Ant carrying away a dead bee

Scout bees performing two types of dances

Honey bees have developed a remarkable form of communication in order to convey information on food sources to their colony mates. Some flying bees act as scouts. They look for rich sources of food and communicate the location of these food sources to the bees in the hive by performing two types of dances on the honeycombs (or, as in the image series, the bodies of other bees) as their dancefloor.

Indicating the direction

For food sources within a distance of 100 m, they perform the round dance, and for greater distances, the so-called waggle dance. To perform the round dance, the bee walks for up to three minutes in small circle, and after completing afull circle, it walks towards the direction of the food source.

Indicating the distance

The more circles the bee completes per unit time, the closer the food source is located to the bee have. The bee's waggle (or lack thereof in the case of the round dance) is another decisive indicator. When the bee performs a waggle dance, the number of abdominal waggle movements and the vibration of the thoracic muscles convey information on the food source's distance as the bee walks straight ahead in figure-eight loops.

The direction in relation to the sun, "fine-tuning" on site

The direction is always indicated in relation to the current position of the sun. Bees can determine the sun's position even under a clouded sky, because they perceive the polarisation pattern of the sky. The Austrian zoologist Karl von Frisch discovered and analysed this form of communication, which earned him a Nobel Prize in 1973. Numerous follow-up studies have since contested whether the coding of the bee's dancing is as precise as described by Frisch, because bees only require approximate indications of the food source's direction and distance and they will determine the exact location through olfactory signals.

The mathematical description of the waggle dance (distance of more than 100 m, sketch): Dancing bees show their sisters where a food source or a suitable location for a new colony is. The central axis of the dance figure signals the direction (the polar angle φ to the subsolar point), and the frequency of the abdominal waggling along this direction is an indicator of the distance r. The image series above shows individual shots, in which a scout bee conveys information on a suitable spot to re-establish its colony. The photographs illustrate how confusing such scenes can appear to the human eye. One can barely discern anything without a slow-motion video (500 shots per second), like the one that can be found on the website www.uni-ak-ac-at/evolution (cf. URL below).

http://sodwana.uni-ak.ac.at/dld/bienentanz.avi
Frisch K.v. **Über die „Sprache" der Bienen, eine tierpsychologische Untersuchung.** *Zoologische Jahrbücher (Physiologie) 40: 1–186 (1923)*
Frisch K.v. **Tanzsprache und Orientierung der Bienen** *Springer, Berlin-Heidelberg-New York, 578 Seiten,*
http:// alotof.org/ wiki/ media/ d/ dc/ Dr._Karl_Von_Frisch_\%28auth.\%29_Tanzsprache_und_Orientierung_der_Bienen_1965.pdf (1965)
Tautz J., Heilmann H.R. **Phänomen Honigbiene** *Springer-Spektrum Heidelberg (2012)*

Not unlike a hummingbird

The hummingbird hawk moth (*Macroglossum stellatarum*) is a diurnal hawk moth. Hawk moths (Sphingidae) are usually nocturnal. The hummingbird hawk moth's name is derived from the fact that it resembles the small New World bird in size and appearance.

30 wingbeats per second

The image series were shot at 240 frames per second; the full wingbeat cycle of a hummingbird hawk moth takes eight times 1/240 seconds, which equals a wingbeat frequency of 30 Hz. This wingbeat frequency is very high for a hawk moth, and is not reached by larger species of the family.

Ideal adaptation to airflow despite chubby body

The insect's chubby trunk is quite striking. A vast number of elongated scales provide a streamlined coating. Such low Reynolds numbers and a tear-shaped, elongated trunk do not lead to an ideal adaptation to the airflow.

The hindwing hovers for a moment at the lower reversal point, while the forewing is folded over and raised to initiate the upstroke.

Childhood memories

Endemic and popular around the world

The semi-spherical ladybird (Coccinellidae) can be found on all continents. They are popular everywhere, one reason being that they are avid exterminators of aphids. We typically think of the most "emblematic representative" of this beetle family, the seven-spot ladybird (large image to the left and image at the top right). What child has not held a ladybird on the tip of their finger? The beetle at the bottom right is an Asian ladybeetle, which has also become endemic to our regions.

Seven-spot ladybird (*Coccinella septempunctata*)

Seven-spot ladybird (*Coccinella septempunctata*)

Unpacking the wings

A short instance of "hesitation" when taking to flight can be observed as the ladybird spreads its large wings, which lie folded beneath the elytra. Then everything happens very fast. An analysis of the take-off phase can be found on p. 140. The unfolding of the wings is achieved by means of a bistable vein mechanism.

Asian ladybeetle (*Harmonia axyridis*)

Take-off!

The number of spots

is, of course, not an indicator of the beetle's age. It is a species-specific characteristic that does not change over the course of a beetle's life. With some ladybird species, there is variation of coloration within the same species, such as red with black dots, but also, though to a rarer extent, black beetles with red dots (melanism). The two beetles pictured below belong, in fact, to the same species, namely the Asian lady beetle (*Harmonia axyridis*), which is incredibly variable in appearance!

A "flip switch" enabling the wing's double unfolding

The shadows in the image series above clearly illustrate that the ladybird's wings are easily twice as long as its elytra. So a simple folding of the wings is not enough to tuck them away. They must be folded twice at least. The unfolding of the wings is achieved by means of a bistable vein mechanism (functioning like a flip switch, which is also stable in two positions), which then leads to a stiffening of the wings' surfaces. See also the illustrations on p. 138.

The elytra are also used during flight

It is often assumed that the elytra's sole purpose is to protect the membranous wings. They are simply spread during flight and then act as aerofoils, which, as measurements have shown, they do quite efficiently. The image series to the left shows, however, that they flap along with the hindwings albeit with a phase shift. That is to say, they actually act as flapping wings, though they only contribute to 15 % of the aerodynamic force.

80 wingbeats per second

Both image series on this page were shot at 240 frames per second. The process pictured above thus lasted about 1/16 of a second and the series to the left twice as long (1/16 second for the left and right column respectively). From this we can deduce a wingbeat rate of about 80.

Tilt stability during slow flight

In several individual shots, it can be seen that the small white (*Pieris rapae*) raises its abdomen slightly. This is relevant for its flight stability, especially during slow flight, when it appears to be hovering. The butterfly must always make sure to "counter-balance" any torque forces. This balance is necessary to ensure tilt stability.

Raising and lowering the abdomen

When the aerodynamic forces act ahead or behind the centre of mass, the flying object is prone to rotations about the lateral axis, head up or head down depending on the current position of the wings. The butterfly's inertia, and especially that of its heavy abdomen, dampens such tilt vibrations, but they are not completely eliminated.

Trimming like fighter planes

When the butterfly raises or lowers its abdomen, it can achieve a sort of trimming and adapt it to the desired flight condition. In the past, fighter planes were equipped with a trim weight, which could be moved forward and backward by means of a crankshaft. The butterfly's raising and lowering of its abdomen functions in the same way.

Flying with dishevelled wings?

The image to the left shows how ragged the trailing edge of a butter-fly's wing can be. This butterfly's wing is only moderately torn; there are even worse cases of insects with torn wings that can still fly. An extreme example of this are large flies that spend a whole season cir-cling the top of a mountain, a popular gathering spot for males and females for the purpose of copulation. After one such season, the trailing edges of a male's wings are typically quite frayed. They can still fly quite well, albeit not as effectively and fast as before. Since the airflow around the wings is no longer optimal, these flies buzz noticeably – a sure sign of aerodynamic stall. Moreover, flying under such conditions costs more metabolic energy than usual.

Conclusive evidence of biological kinship

Within the clade Neuropterida, we can find the order of net-winged insects (Neuroptera, once known as Planipennia). The most widely-known net-winged insects are the green lacewings (Chrysopidae). Some *Chrysoperla* species have golden eyes, which is why lacewings are also called "gold eye" (Goldaugen) in German. Antlions (Myrmeleonidae) are another group of net-winged insect species. Their larvae are known as doodlebugs due to the cone-shaped marks they leave in the soil.

Green lacewings: hearing and being heard ...

Green lacewings are a group of insect species that are particularly striking in the bright-green or brownish colour of their large wings. Due to the size of their wings, they are quite cumbersome in flight. They are mostly nocturnal, and like many moths, they have an auditory organ that enables them to capture the ultrasonic calls of bats in order to let themselves drop quickly and escape on time. Some species produce soft chirping sounds, which are used as species-recognition signals during courtship.

... and cunning disguise

Imagines and larvae are avid leaf-eaters. The larvae of green lacewings bear hooked bristles on their backs, which are provided so that they can throw the remains of aphids they have sucked dry to the back of the larva's body with their long mandibles, where these dead aphids remain stuck as camouflage. In this manner, the larvae of green lacewings manage to catch aphids from colonies that are guarded by ants without being recognized.

Aspöck, U., Haring, E. Aspöck, H. **The phylogeny of the Neuropterida:**
    long lasting and current controversies and challenges (Insecta: Endopterygota)
    Arthropod Systematics & Phylogeny 70 (2): 119-129 (2012)
Aspöck, U., Aspöck, H. **Kamelhälse, Schlammfliegen, Ameisenlöwen ... Wer sind sie?**
    (Insecta: Neuropterida: Raphidioptera, Megaloptera, Neuroptera)
    Stapfia (Linz) 60: 1-34 (1999)

The animal pictured here is a *Nothochrysa fulviceps*, a large, widespread, but rather rare species. You can see well the fine-mesh, net-like veins of the wings, which earned this whole group of insects its name. Some of these species can be found in our homes during autumn, where they look for a safe winter habitat.

Unmistakable with their long necks

Snakeflies (Raphidioptera) are an order of net-winged insects comprising about 225 species worldwide (though only on the northern hemisphere). The genera within this order are divided into two families. These insects are distinguished by their elongated prothorax (neck) and their wings that are folded over the abdomen, which lend them an unmistakable appearance.

On their way, whirring …

All snakeflies are terrestrial in all stages. The imagines are diurnal and not good at flying, so they usually just walk or travel a few centimetres in flight, whirring close to vegetation. Some species (especially those whose larvae develop in the soil) prefer to dwell on low to moderately high scrub vegetation. The members of the family Raphidiidae are all predators that primarily feed on aphids; the imagines of the rarer family Inocelliidae feed mostly on pollen. The animal pictured here is a Xanthostigma xanthostigma, a species that is more common in our regions. The long ovipositor at the rear of its abdomen indicates that it is a female.

Charles S. H. **Acoustical Communication during Courtship and Mating in the Green Lacewing *Chrysopa carnea* (Neuroptera: Chrysopidae)** *Annals of the Entomological Society of America 72 (1): 68-79 (1979)*
Aspöck, H., Aspöck, U., Rausch, H. **Die Raphidiopteren der Erde, 2 Bände** *Goecke & Evers, Krefeld, ISBN 3-931374-27-0 (1991)*

Enormous jumping power

Strained to breaking point

The jumping force of grasshoppers is enormous. The leverage ratios of its hind legs are optimally developed for this purpose. The powerful jumping muscles of the thick femur (upper thigh) contract only for a short moment, but they do so with great force. In the process, the ligaments are strained almost to the point of rupture. Technically speaking: Taking the large desert locus *Schistocerca gregaria* as an example, the factor of safety is only about 1.25.

10 g acceleration!

At the start of the jump, the thin and thus light tibia supports itself on the ground. The initial acute angle between the femur and the tibia is stretched to almost 180° within a few milliseconds. The grasshopper then launches itself up into air diagonally. The desert locus, in fact, accelerates with a force that corresponds to 10 times the acceleration of gravity. They can jump a distance of a metre. During the jump, the flight machinery is fired up.

Tremendous acceleration values in the animal kingdom

Click beetles (Elateridae) and jumping plant-lice (Psylloidea) are typical examples of so-called "explosion jumpers". The soldiers of some termite species have hooked mandibles that cross over each other. With the enormous centrifugal force that is released upon snapping open their mandibles, these termites can easily tear up an ant. Similarly astonishing cases can be observed in the kingdom of plants, as shown on the right-hand page.

Brown R. H. Mechanism of locust jumping *Nature (Ldn) 214, 939 (1967)*
Scott, J. The locust jump: an integrated laboratory investigation. *Advances in Physiology Education Published 1 March 2005 Vol. 29 no. 1, 21-26 DOI: 10.1152/advan.00037 (2004)*

20 bar "vanish" within a millisecond

The fruits of all species comprised by the genus *Impatiens* develop great internal pressure while ripening. A sudden equalization of the pressure causes the fruit to explode and dehisce into five distinct units. During the explosion, the seeds are scattered at very high velocities.

One of the fastest mechanisms in the plant kingdom

The fruits are specially designed for the realization of this mechanism. The five septa of the seed capsule contain expansion tissue producing turgor pressure, which swells against the resistance of the thick pericarp and thus generates internal stress. During the ripening process, the middle lamella between the septa break down and turn into increasingly thin and weak cell walls, allowing the high pressure inside the fruit to be eventually released. The cell walls burst, and the septa roll inward forcefully. During the first millisecond of the explosion, they turn into a kind of pouch that encloses the seeds. This pouch is then tied up forcefully, which causes the seed weighing 300-4000 g (!) inside to be released and scattered.

Great flight distances of 5 metres and more

Having been sent out into the world, the seed can fly over 5 m on a horizontal plane and even further on sloping ground. This explosion mechanism thus permits a fast distribution of the plant's seeds.

Wolters, B. *Impatiens parviflora* (Balsaminaceae). Aufspringen der reifen Frucht (Turgormechanismus) *Encyclopedia Cinematographica E 723/1964, IWF Göttingen (1966)*
Nachtigall, W. Fruchtexplosion und Samenausschleudern beim Kleinblütigen Springkraut *Impatiens parviflora*. **Teil 1: Bau und Funktion der Frucht,**
Teil 2: Stroboskopische Messungen und Rechnungen zum Samenausschleudern *Mikrokosmos 2010;99/4: 211-217 und 99/5: 296-302 (2010)*

Flying ants

Females exclusively – males only on demand

Together with wasps and bees, ants are insects of the order Hymenoptera. Within the subclade Aculeata (the females of which have an ovipositor that is modified into a stinger), they form a family of over 13,000 species worldwide. All species are colonial (eusocial), and they display a division of labour that can be very complex. The colonies consist of female animals exclusively, comprising one or more queens, workers, and different kinds of soldiers. Sexual ants, that is, males and new queens, are only rarely produced for the purpose of mating.

Thieves, seed eaters, or fungus cultivators

The colonies may consist of a few hundred or even up to several million individuals, depending on the specific species. Although most ants can sting, the stinger has been lost or reduced in some groups, such as the wood ants, in favour of other forms of defence, such as biting and the spraying of formic acid. The majority of ants are thieves, some are seed eaters, and some, such as the tropical leafcutter ants, are fungus cultivators, which carry around pieces of leaves on which fungi can grow. These ants feed on the fruiting body of the cultivated fungus.

Males of a South African species of trap-jaw ants Odontomachus, which are well-known for the snapping mechanism of their mandibles. The mandibles can snap shut at an enormous velocity, literally tearing the prey into pieces.

The males of most species – pictured here a specimen from the diverse group Formicinae – are considerably smaller than their queens, as their sole purpose is limited to mating. After copulation, the males die, whereas the queens begin to search for new nest sites.

Larabee F.J., Suarez A.V. **Mandible-Powered Escape Jumps in Trap-Jaw Ants Increase Survival Rates during Predator Prey Encounters**
PLOS ONE | DOI:10.1371/journal.pone.0124871 May 13, 2015, 1-10 (2015)

Brilliant form of communication via scent

All colonies are highly organized, and work with a remarkable form of communication that is based on scent and allows to solve nearly all problems within and outside of the colony. All workers can produce trail pheromones that can be traced by other individuals in order to find their way back to the nest, aggregation pheromones to mark prey, alarm pheromones when a concerted defence is needed, nest-specific pheromones to distinguish nest-mates from outsiders, etc.

Only males and queens can fly

The mating flights of many ant species are quite spectacular. While the female workers are wingless, sexual ants such as males and new queens are born with wings. Many thousands or even millions of males and potential queens often leave their nests simultaneously in order to engage in a nuptial flight. Male ants thus find queens from other nests that are ready for copulation.

Copulation often occurs in flight,

with the much larger queen ant simply dragging along the smaller male ant. After copulation, the males die, and the queens lose their wings and start looking for new nest sites.

Giant superorganisms

The queen begins to lay a small batch of eggs, from which larvae hatch. These larvae, which still need to be fed by their mother, then develop into adult workers.

The foundation of new nests can take various forms. Fertilized queens may also intrude into already existing nests and be adopted by these colonies. Such nests have several queens. Over the years, these colonies can develop enormous populations. In addition, old queens may emigrate with parts of the population and construct auxiliary nests. Such nest clusters comprising numerous auxiliary nests can extend over thousands of kilometres, forming giant superorganisms.

Hölldobler B., Wilson E.O. **The Ants** *Harvard Univ Press (1990)*
Hölldobler B., Wilson E.O. **The Superorganism: The Beauty, Elegance, and Strangeness of Insect Societies** *Norton & Company ()*

Phenomenal sensors

Cockroaches – pictured here an American cockroach (*Periplaneta americana*) – are not held in particularly high esteem, even though they are insects of very great interest. The larger species are fast runners and can reach velocities of up to 1 metre per second (equivalent to walking pace). When in imminent danger, they prefer to hide in cracks. For this purpose, they are equipped with various sensors. Their tactile sense, which they obtain from sensory hairs, is remarkably sophisticated. It enables these insects to detect substrate vibrations of atomic dimensions (!) and to escape from a dangerous situation quickly. They do not like to be discovered, because it costs them much metabolic energy to take to flight.

With loud buzzing

When they are discovered, cockroaches make a loud buzzing noise. Their two pairs of wings are differently developed. The forewings are narrower and more heavily sclerotized; the hindwings are thinner and fan-like. While at rest, the forewings are placed like protective elytra over the folded hindwings.

Fore- and hindwings beat out of phase

Left-hand image: The cockroach's forewings have reached the upper turning point, initiating the downstroke. The hindwings have reached the lower turning points and are about to beat upwards.

Right-hand image: The forewing on the left-hand side clearly shows the characteristic twisting of the cockroach's wings. For more details, see p. 199f.

American cockroach (*Periplaneta americana*) with an egg case sticking out from the abdomen.

Flying termites

Termites are actually cockroaches – with a king and queens!

More recent phylogenetic studies have classified termites as a family of cockroaches. Similar to ants, they are all eusocial, but as opposed to all colonial species of Hymenoptera, such as our honeybee or colonial wasp species, they also have male workers and a king. Due to the white to pale yellow colour of most species, they are often called "white ants." Termites can be found around the world, but most of the over 3000 species are endemic to tropical regions. In Europe, only 10 species are known to live in the Mediterranean area.

Digestion of cellulose by means of symbionts

Termites feed exclusively on plants and cellulose material such as wood, which they can digest with the help of symbionts inside their intestine. This is why they are dreaded as pests to wooden buildings in the tropics.

engage in nuptial flights. These nuptial flights typically occur at night. Termites are not very skilled at flying because they struggle to manoeuvre with their elongated wings, which are considerably longer than their body. They can easily be observed against the light, to which termites are also attracted (see image at the top right).

Ants are their greatest enemies

Under the light, termites may encounter all kinds of enemies, such as ants, which are their greatest predators (see image at the bottom right). Future queens and kings may also find each other in this manner, though. Once they've found each other, they shed their wings and seek out a suitable site for a new nest.

Mating occurs in the new nest

It is only once they have established a new nest, that the queen and king termite mate. The queen then lays a small batch of eggs, from

Giant colonies with millions of individuals

The colonies of termites, similar to those of ants, can grow to remarkable proportions and comprise several million individuals. Most termites are between 5 mm and 15 mm in length. Their queens, however, can reach a size of up to 14 cm and live as long as 40 years.

Three castes

The individuals are divided into three castes: Sexual animals, workers (sexually undeveloped males and females), and soldiers. Termites can build mounds from earth and cellulose material that are quite remarkable in size. In African and Australian savannas, these termite mounds are sometimes several metres tall.

Only sexual animals have wings

As with ants, only sexual termites grow wings, with which they can

which nymphs hatch after a short while. These nymphs already look like small termites and they immediately start to carry out their work on the nest.

Termites do not have larvae

Termites belong to the orders of insects that undergo incomplete metamorphosis (Hemimetabola). That is to say, they do not have a pupal stage as larvae, but already hatch as miniature versions of the adult insect. As they moult their skin, the insects grow and gradually develop their wings (unless they are wingless). Once the colony has reached a minimum size, the queen may lay up to a thousand eggs a day. King and queen termites never leave their nests.

Ware J.L., Litman J, Klass K.D., Spearman L.A. **Relationships among the major lineages of Dictyoptera: the effect of outgroup selection on dictyopteran tree topology**
Systematic Entomology 33 (3): 429–450 (2008)
Bignell D.E., Roisin Y., Lo N. **Biology of Termites: A Modern Synthesis**
Dordrecht: Springer. ISBN 978-90-481-3977-4 (2010)

Fast predators in the same habitat

Dangerous beaches

The predatory tiger beetle *Cicindela* (*Calomera*) *littoralis* primarily preys on kelp flies and small sea slaters on sandy seashores. The diurnal predators, which can see very well in three dimensions due to their prominent compound eyes, simply run down their potential prey.

Fast take-off – when it is warm enough

As opposed to normal beetles, tiger beetles can take to flight very fast when they are disturbed (right-hand page at the top), and they will usually land again after only a few metres. However, in order to achieve such a fast take-off, the weather must be warm, because these

insects are cold-blooded and their muscular activity is temperature-dependent. Tiger beetles do not raise their elytra at an angle, and neither do they flap them like cockchafers (cf. p. 242). Instead, they spread their elytra to the side and hold them with their front pairs of legs. In that manner, the tiger beetle's elytra function like the aerofoils of high-wing aeroplanes. The whole take-off process lasts no more than 0.12 s, and during its short flights, the tiger beetle's velocity does not exceed 3 m/s.

Another "dashing lightning bolt"

Other dangerous predators live in the same habitat: sand wasps from the digger wasp family. The male wasps look for females, while the females prey on flies that they can "serve" to their larvae.

Nachtigall, W. **Take-off and flight behaviour of the tiger-beetle species** *Cicindela hybrida* **in a hot environment** *Entomol. Gener. 20 (4), 249 - 262 (1995)*

Searching for females ...

The sand wasp *Bembix rostrata* is "looking for a wife". The tiger beetle is comparable in size and stature to the female wasp, but it is too large to serve as prey. These *multiple images* shot at a rate of one twelfth of a second show how the wasp (1) approaches the beetle, (2) "examines" it, and (3) turns around to fly away.

Hemelytra

Differently shaped pairs of wings

The dock bug is a species of true bug (Heteroptera) that is common in our regions and belongs to the family of leaf-footed bugs (Coreidae). The wings of these bugs are narrower than their abdomen, which is rounded upward at the edges. This species likes to dwell on low vegetation and suck its sap. For this purpose, all true bugs are equipped with elongated mouthparts of the piercing-sucking type, which is why they were formerly known as Rynchota. Their wings are placed over the abdomen, lying flat on top of each other. The anterior portion of the wings, known as corium, is hardened, whereas the apical portion is membranous. Such wings are also known as hemelytra, and that is also what the name of this insect order, Heteroptera, refers to. The basal half of the forewings is heavily chitinized, resembling elytra, but they gradually taper to fine tips. The hindwings, on the other hand, are fully membranous. While the bug is at rest, the hindwings are folded longitudinally and hidden beneath the protective cover of the hemelytra.

Dock bug (*Coreus marginatus*)
Preparations before the bug takes to flight can take quite long (about a second)

Fore- and hindwings are coupled

Fore- and hindwings are linked by means of a sophisticated skid mechanism so that both wings are firmly coupled during flight. As a result, the wings move together on upstroke and downstroke, like a single unit. In the photograph above, as well as in the picture of a minstrel bug below, this wing coupling can clearly be seen. Before the insect takes to flight, the forewings and hindwings are coupled by means of a folding mechanism to form a single large wing surface. The picture of the dock bug at the bottom left shows how the forewings are already spread as the hindwings are about to unfold.

The abdomen's red colour is only visible during flight

As the insect takes to flight, it reveals the bright red upperside of its abdomen, which is usually covered by the wings. The bright red colour is possibly supposed to ward off predating birds.

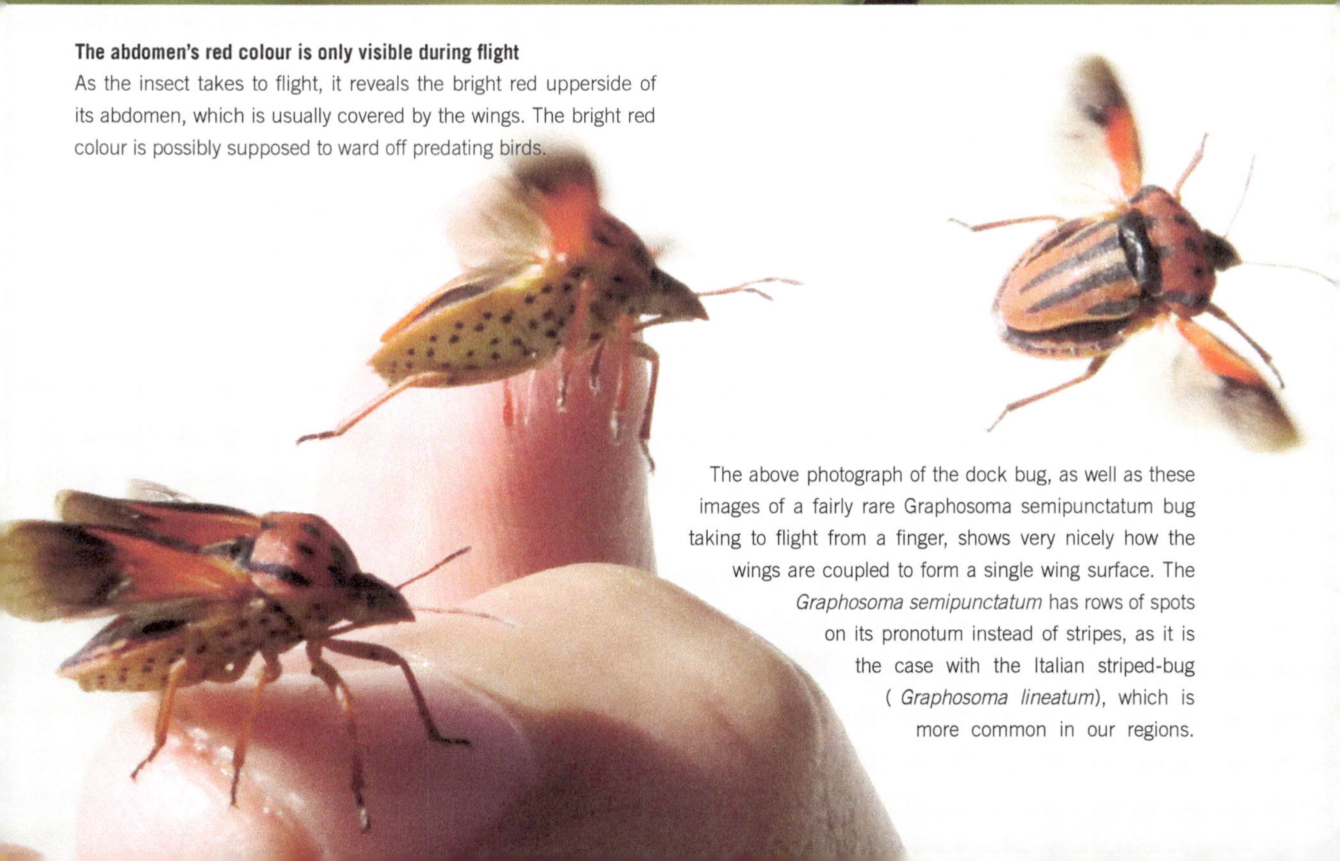

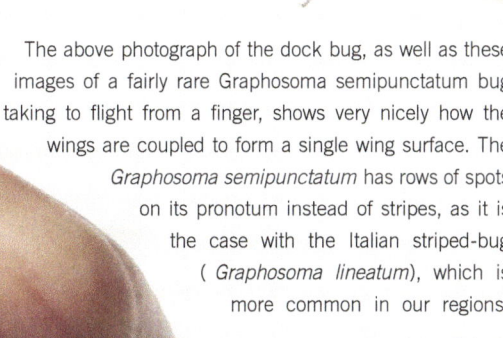

The above photograph of the dock bug, as well as these images of a fairly rare Graphosoma semipunctatum bug taking to flight from a finger, shows very nicely how the wings are coupled to form a single wing surface. The *Graphosoma semipunctatum* has rows of spots on its pronotum instead of stripes, as it is the case with the Italian striped-bug (*Graphosoma lineatum*), which is more common in our regions.

Stationary hovering flight

Furry bee (*Anthophora aestivalis*)

Among the wild bees, the species of the furry bee genus *Anthophora* are particularly good fliers. They are extraordinarily fast and can make rapid turning manoeuvres, remain in hovering flight on the same spot for several seconds, and examine their surroundings in the process. The females are avid pollen collectors and feed their offspring in ground nests that they have dug themselves.

Marmalade hoverfly (*Episyrphus balteatus*)

"As if nailed to the ground"

Male hoverflies – pictured here a marmalade hoverfly *Episyrphus balteatus* – like to stand in the fingers of light that are produced on the ground by the sun shining through the branches of trees. They hover at high frequencies of about 300 Hz and more but only with low amplitudes.

No conclusive evidence about how this is possible!

During flight, their body is kept horizontal, and the wings beat at an inclined angle. There is no conclusive evidence yet about how this is possible; usually the very opposite is the case: see, for instance, the stationary hovering flight of birds.

Acceleration at 7 g

When an object passes them at a high velocity, hoverflies rapidly accelerate in an attempt to catch up, reaching peak acceleration rates of about 7 g. If they are passed by a female, male hoverflies will chase it in order to achieve copulation. Otherwise they will just as quickly return to their resting position. These rapid acceleration manoeuvres can be triggered by throwing tiny stones or pieces of bread past these insects. Apparently, it is only at a close distance that hoverflies can discern the objects that they have been chasing after.

Dumbo's flight to the sun

Proportions

The Colorado potato beetle has the figure of a large ladybird. Reaching a length of one centimetre, the potato beetle easily weighs four to eight times as much as a ladybird, but they still perform in a similar league (a stag beetles weighs 200 to 300 times as much).

An elephant cannot jump, let alone fly. This is simply due to its size. An elephant measures 400 times the length of one potato beetle, and it takes approximately 50 or 100 million specimens of this tiny beetle to equal the weight of one elephant. Even the biggest ears cannot make up for this (see right-hand page, below).

Take-off towards the sun

Once the elytra are opened (bottom left), the carefully folded wings are spread and the insect gets ready for take-off. As can be gathered from the position of the shadow, the Colorado potato beetle typically takes to flight in the direction of the sun. Like a ladybird, the beetle draws up its hindlegs for take-off.

Colorado potato beetle (*Leptinotarsa decemlineata*)

161

Navigating by the position of the sun

The question as to why many insects fly in the direction of the sun has already been dealt with on pp. 64f.: These animals find their way using the sun or the polarized light in the sky, which is also determined by the position of the sun (even as it hides behind a cover of clouds).

Advantage for photographers …

Photographers can expect insects to take to flight in the direction of the sun. So while taking photographs of insects, this knowledge can be used to position the camera in a manner that enables them to obtain "optimum focus" for as long a duration as possible.

The two image series to the right show an Asian (they have been introduced to Europe by now) ladybeetle (*Harmonia axyridis*) and an ichneumon wasp as they take to flight. The shadows cast by the insects provide some hints about the position of their wings in relation to the flight direction. If the insects take off in the direction of the sun, the shadows cast by the wings appear symmetrical. Both image series could only be shot using a very short exposure time (1/6400 s) in order to "freeze" the motion of the wings.

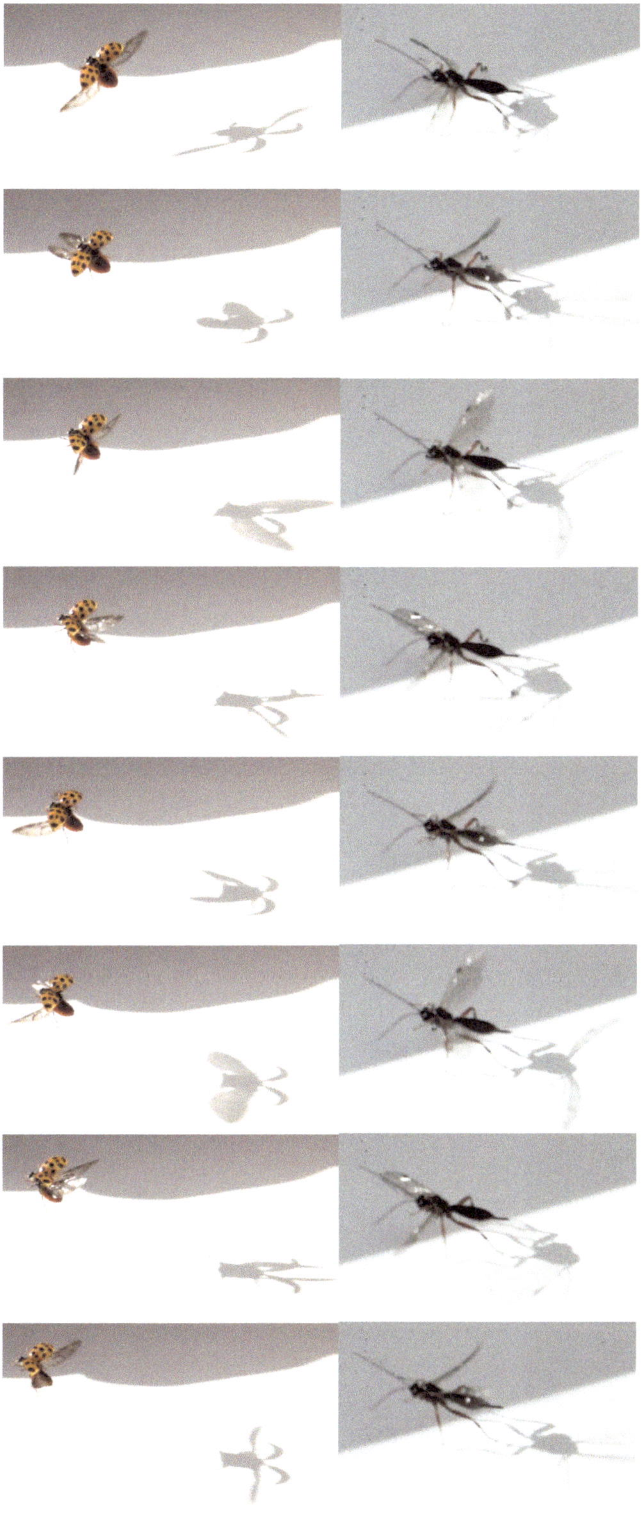

Analysis of a butterfly's flight

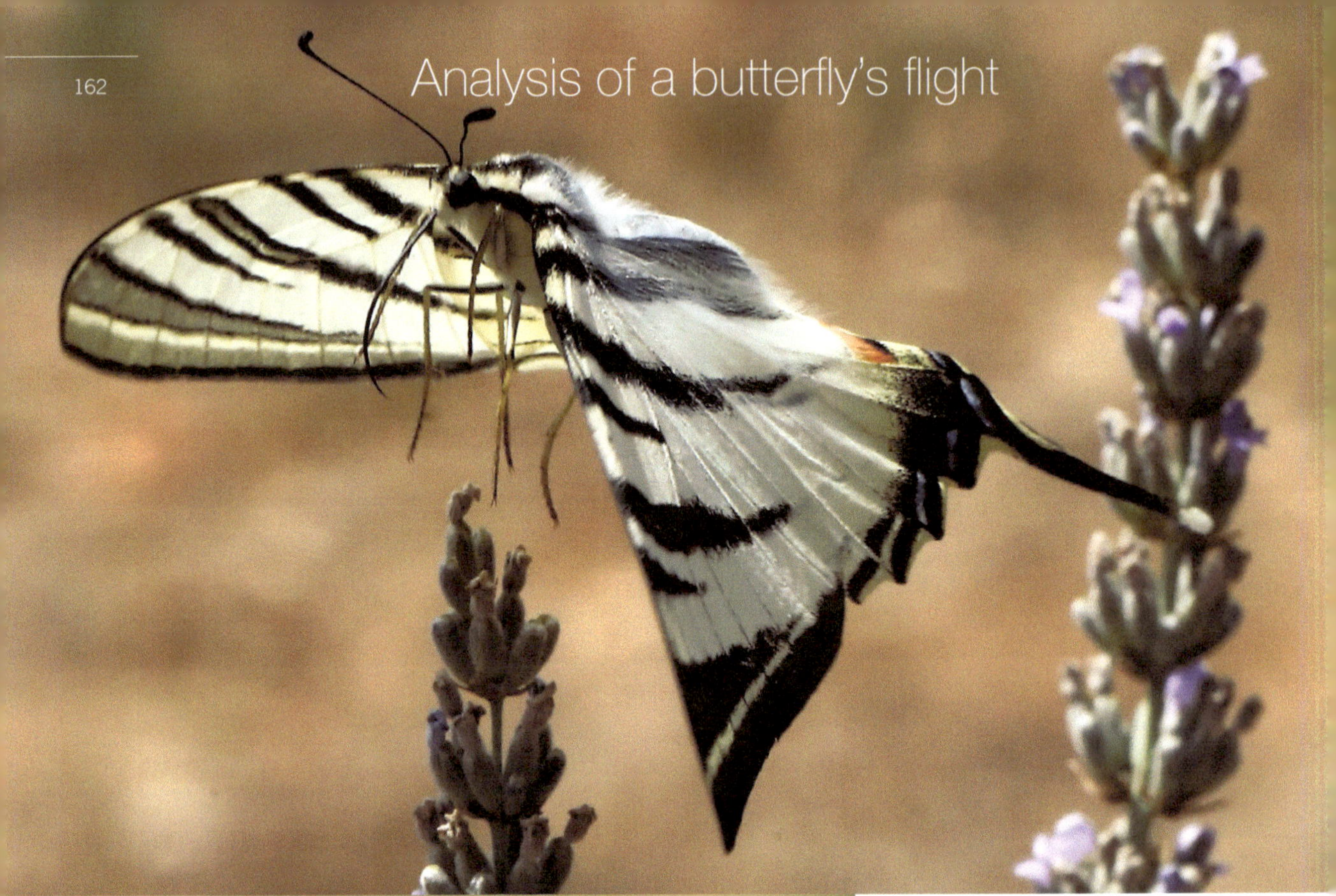

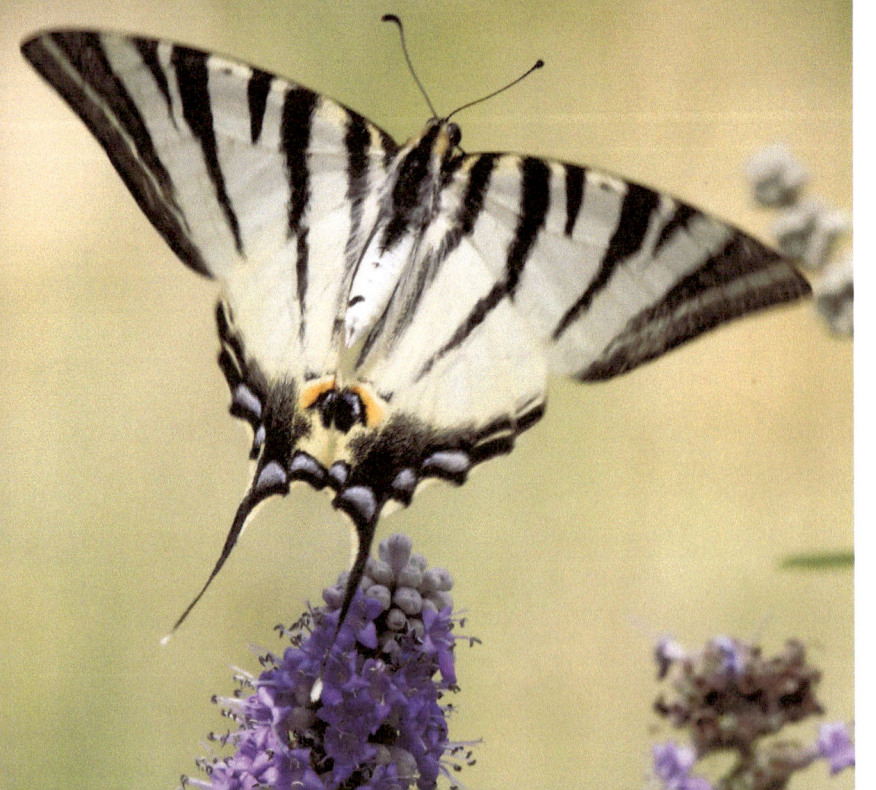

Scarce swallowtails (*Iphiclides podalirius*), see also p. 243, can catch updraughts to soar for longer periods without a wingbeat.

Right-hand page: A small tortoiseshell (*Aglais urticae*) takes to flight and makes a quarter-turn rotation with one single wingbeat (image 1 to image 7). Since the images were taken at a rate of 1/250 s, this equals a duration of about 1/40 s. Slight differences between the left and right pair of wings are recognizable on close inspection. The left pair is significantly more curved than the right pair, its leading edge is pulled down more than the other, and it appears to be slightly ahead of the right pair. More lift is generated on the left side than on the right, which causes the butterfly to rotate around its longitudinal axis. The rotating butterfly slightly overshoots but eventually manages to steady itself.

6 Birds: The "classics" among flying animals

From the hummingbird
to the Andean condor

Among the vertebrates, the ubiquitous birds are the epitome of
flying. The Archaeopteryx – considered a classic "missing link"
in evolutionary theory – and its immediate ancestors may be taken as the starting point of all bird life 200 million years ago.
Since then, birds have been flying, whirring, and gliding in all
sizes, pollinating flowers or preying on insects and small mammals.

Archaeopteryx

Visible reptile features:

1 = long tail with vertebrae

2 = beak with peg-shaped teeth

3 = fingers with claws

Bird features:

4 = asymmetrical flight feathers

5 = reversed first toe

Further reptile features:

• small and inconspicuous cranial vault

• column with no fusion of vertebrae

• ribs without uncinated processes

• gastralia present

• pelvic bones only bound together by connective tissue

• separate metacarpals and fingers with claws

• tibia and fibula are not fused

Further bird features:

• bird-like arm skeleton

• presence of a furcular (fused clavicles)

• bird-like pelvis

• leg skeleton similar to that of cursorial birds

• sternum to anchor flight muscles (though still very small)

• only partially fused metatarsals

Birds are descended from dinosaurs

It has been known for quite a while that birds are related to reptiles and that they must actually have evolved from them. This has been supported by findings of the famous primeval bird *Archaeopteryx*, a kind of evolutionary link between reptiles and modern birds.

Fossilized feathers and then a whole skeleton

A fossilized feather was found in 1861, preserved in Solnhofen limestone in southern Germany, followed by the discovery of a nearly-complete skeleton three years later. Since this discovery took place only shortly after the publication of Darwin's theory of natural selection, the *Archaeopteryx* has become a kind of icon of palaeontology and evolutionary biology.

Eleven specimens of *Archaeopteryx* have since been found in different states of preservation and completeness, mostly in the region of Eichstätt/Solnhofen. Depending on the location where the fossil remains are currently housed, they are known as the Berlin specimen (Museum für Naturkunde), the London specimen or the Solnhofen specimen.

Reptile features and bird features

A detailed analysis of the *Archaeopteryx*'s many features has shown that it "still" exhibited numerous reptile features (such as jaw with teeth, free fingers with claws on their wings, tail with vertebrae), but it was already showing many features of modern birds: a bird's skull, plumage, a shoulder girdle with a coracoid to support the flight muscles, the shape of its shortened arm bones (wings), bird legs (the metatarsal is connected to the walking leg), bird feet (the first toe is reversed).

Archaeopteryx could certainly fly

Archaeopteryx lived some 150 million years ago, that is, towards the end of the Jurassic, and it certainly had the capacity for flight. This could be inferred from a series of morphological features. Like modern birds, the *Archaeopteryx* had a lightweight skeleton, consisting of bones with air pockets. The flight feathers were asymmetrically shaped, meaning that the vanes to the left and right of the rachis differ in length. So, as with modern birds, it is possible to distinguish between the outer vane and the inner vane of a feather.

Godefroit P., Cau A., Hu D-Y., Escuillié F. , Wu W., Dyke G. **A Jurassic avialan dinosaur from China resolves the early phylogenetic history of birds** *Nature. 498 (7454): 359–362. (2013)*

Alonso P. D., Milner A. C., Ketcham R. A., Cookson M. J., Rowe T. B. **The avian nature of the brain and inner ear of Archaeopteryx** *Nature 430 (7000): 666–669 (2004)*

This aerodynamic shape of the feathers can only be found in birds that are active flyers. The feathers of cursorial birds, such as the ostrich or emu, have become symmetrical (again).

The first "missing link"

This primeval bird provided evidence in support of Darwin and his theory of natural selection, proving that the missing links predicted by Darwin's theory actually existed. Accordingly, *Archaeopteryx* may be defined as a primeval bird species living during the Jurassic that represents a transitional form between reptiles and the true birds living today.

Everything comes to light in the end ...

The London specimen, which was discovered in 1861 at Langenaltheimer Haardt near Solnhofen, is one of the three most important specimens. It was the first finding of a complete skeleton, and is the type specimen of the species *Archaeopteryx lithographica*. Only a few months after its discovery, the specimen was purchased by the British Museum under the instructions of Richard Owen, who was the head of the Museum's collection of natural history and a declared opponent of Darwin's theories. By purchasing the skeleton of the primeval bird, Owen wanted to hide evidence supporting Darwin's evolutionary theory. The fossil was kept under lock and key for a long time, and the results of its examination were only gradually published in small portions.

Archaeopteryx lithographica, the "London specimen" with particularly well preserved feathers.

Some features of modern birds

1 = tail without vertebrae
2 = bird skull with big eyes and beak without teeth
3 = asymmetrical flight feathers
4 = bird legs (co-ossified metatarsals that are fused to the walking leg)
5 = bird feet (reversed first toe)
• shoulder girdle with coracoid to support flight muscles
• large sternum
• shape of the arm bones (wings)

Pushing the limits of biophysics

How large may the prey be?

This Caspian gull pushes the limits of its flight capacity by dragging an eel to some place where it can devour its prey quietly. The largest European gull, the great black-beaked gull (*Larus marinus*), reaches a weight of 2.22 kg at a body length of 74 cm. The smaller Caspian gull (*Larus cachinnans*) pictured on this page is about 56 cm long and may be estimated to have a weight of about 1.5 kg.

A young male eel can barely be transported

A large female specimen of the European eel (*Anguilla Anguilla*) can, in extreme cases, reach a length of 1.5 m and a weight of about 6 kg. Such a load cannot be carried by a gull. Male eels are only 40-60 cm long and thus considerably lighter, with an estimated average of about 1 kg. So the adult gull that is pictured on this page as it drags away a young and/or male eel can be assumed to carry a load that equals 30% of its own body weight. The gull could possibly carry an additional 50% of weight and even rise into air with this load for a very short period of time. In these pictures, it might already be pushing its limits.

Caspian gull (*Larus cachinnans*)

Caspian gull (*Larus cachinnans*)

On the verge of an airflow collapse

Close inspection reveals that the upper surface of the gull's wing rolls up at the trailing edge – an irrefutable sign that the gull is flying at an extremely steep angle for maximum lift generation, nearly leading to a collapse of airflow on the wing's upper side. The "swerving motion" of the wing's distal end during upstroke is also indicative of the "almost excessive" strain induced by this flight condition. The biological construct "gull" is pushing its physical limits here.

Mere hundredths of a second determine success or failure

The two gulls pictured here are too slow to catch any sardines jumping out of the water. However, this whole process unfolds within mere hundredths of a second. So the reaction time of gulls must lie within this range. That is to say: Before the gull can actually snap at a sardine, the fish has already vanished under water.

Rising contour | Falling contour

$0,5L$

L

Black-headed gull (*Larus ridibundus*) in winter plumage

A gentle introduction to fluid mechanics

Photographs of birds are often astonishingl informative, and with some functional interpretation they can provide much more insight than one migh expect after a fleeting look. Can photographs of a bird' flight be interpreted from the perspective of flight biophysic without delving deeper into fluid mechanics? Comparisons o photographs taken of flying gulls and other birds shall be use as a starting point to discuss and shed light on nine typical an frequently documented aspects.

(1) Contouring of the bird's shape through its plumage

When looking at an image of a bird photograph from the side, th following is revealed: The contour of the bird's trunk widens gradually an continuously after its rounded facial disk. The maximum trunk thicknes does not occur in the first third of the body, as one would expect whe thinking of a teardrop profile, but relatively far to the rear, at about 40 50% of the body length L. The trunk is, therefore, said to have a location o maximum thickness of 0.4-0.5 L. In addition, the trunk's plumage provides streamlined coating to the body. Such a configuration of the trunk produces little drag

Nachtigall W. **Zur biomechanischen Interpretation von Vogelflug-Aufnahmen**
VOGELWELT 135: 83 – 88 (2014)
Tucker V. **Body drag, feather drag, and interference drag of the mounted strut in
a peregrine falcon** *Falco (1990)*

(2) Wing positioning and aerodynamic force generation

Shots that capture a bird's wing on downstroke, as it reaches the middle of its course from the rear top to the front bottom, allow us to sketch those components of aerodynamic force – given stationary airflow conditions – that apply at the moment the photograph is taken. If the wing were perpendicular to the paper, that is, if it protruded out of the book towards the eye of the reader, then all components of aerodynamic force would run flat across the plane of paper so that they could be drawn. Like every body that is enveloped by flow, the wing produces a drag force F_W that works in the direction of the flow. A flat body that is positioned at an angle of attack α to the flow also produces a lateral force F_A that acts perpendicular to the direction of the flow. This force is also referred to, rather misleadingly, as lift, and with good wings, it can be several times greater (in the case of a dove, for instance, about 8 times greater) than the drag force. For the sake of clarity, it been drawn as being only three times greater in the sketch. The aerodynamic force F_{res} which results from these two force components inclines forward at an angle and is thus divided by another parallelogram of forces into a raising force component F_H, and a forward-driving force component F_V, i.e. propulsion or thrust. Lift and thrust are beneficial aerodynamic forces that the bird requires to fly a certain distance. Most wing flap positions produce such beneficial force components.

White stork (*Ciconia ciconia*)

Carrol L. Henderson **Birds in Flight: The Art and Science of How Birds Fly** *Voyageur Press (2008)*
David Goodnow **How Birds Fly** *Periwinkle Books (1992)*

(3) Wings with increasing twisting from inside to outside

Photographs taken of the middle wing section from a certain perspective show that the wing is increasingly twisted from the base to the tip; towards the tip, the leading edge of the wing is increasingly pulled downwards (pronated). A similar twisting can be found with aircraft propeller. The sketch illustrates the impact of this wing twisting.

A wing section can only produce beneficial lift and thrust if the resulting aerodynamic force is inclined forward. This requires a favourable angle of attack α_1, that is, an angle of attack 1 that is not too large, as it may be the case with the proximal portion of the wing. If the wing were not twisted in this way, the distal portion of the wing would be hit by flow at such a large angle α_2 that the airflow would collapse and this section of the wing would fail to produce any aerodynamic force (this is because, due to the constant angular velocity of the flapping wings, the absolute flapping velocity v_{flap} rises as the distance from the wing basis increases, while the flight velocity v_{flight} remains steady; see the two velocity triangles). This negative effect can completely by compensated by means of a pronating wing twisting. The favourable small angle α_1 is thus retained across the whole wing span. This is also illustrated by the sketch on p. xi.

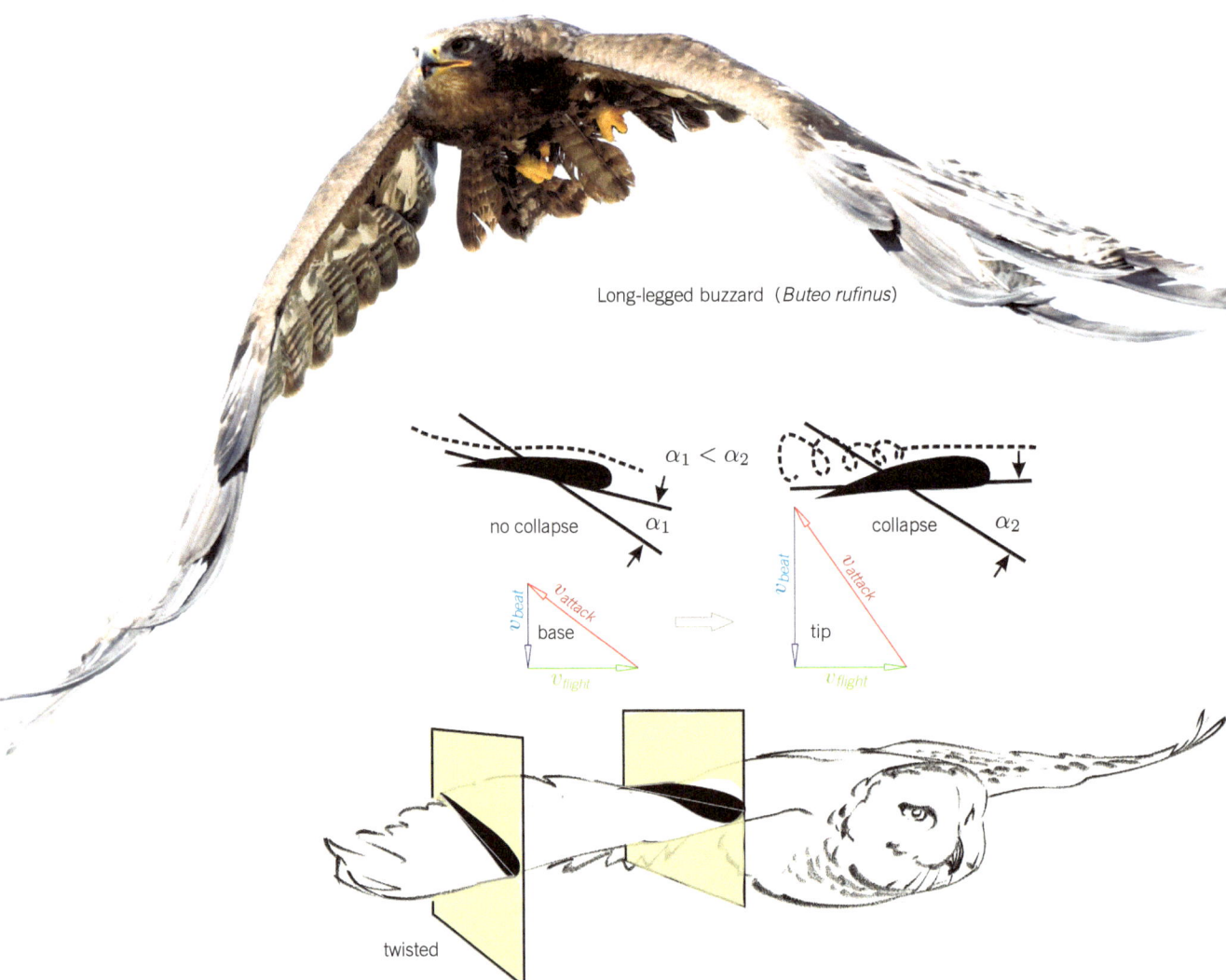

Long-legged buzzard (*Buteo rufinus*)

$\alpha_1 < \alpha_2$

no collapse α_1

collapse α_2

v_{beat} v_{attack} base v_{flight}

v_{beat} v_{attack} tip v_{flight}

twisted

Bilo D., Nachtigall W. **Biophysics of bird flight: questions and results** In: Nachtigall, W. (Hrsg.): Bewegungsphysiologie – Biomechanik. Fortschritte der Zoologie (1977)

(4) Staggered primary feathers

Every single free primary feather generates a short vortex spool. If the primary feathers are staggered conveniently, the individual spools combine to form a tube of vortex spools. Studies have shown that there is a greater dynamic pressure p inside this tube than outside; this arrangement of the primary feathers thus functions as if each tip of the wing were equipped with a jet engine, contributing

How storks became skilled gliders

Latest measurements have confirmed that the white stork will already spread its cascade of primary feathers aeroelastically starting at speeds of about 7 m/s, adjusting for an optimal staggering with an optimal degree of spreading. In this configuration, the stork's wing will produce up to 20% less induced drag than a technical elliptical wing with an elliptical lift distribution, which is known for producing the minimum induced drag within the realm of technology. In this case, nature beats technology, but only if we consider the individual wing. Similar effects can be observed in technical aircraft such as biplanes and triplanes, as well as in wingtip configuration of modern commercial aircraft. The stork thus reduces its sinking velocity, which is particularly important when riding weak thermals: The white stork is geared towards sailing and gliding; its muscles are too weak for flapping flights of longer durations.

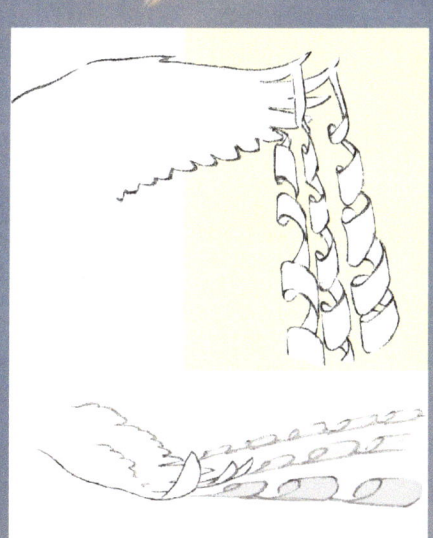

to the generation of thrust. In that manner, the considerably high lift-induced drag at the tip of the wings is somewhat reduced, which saves flight performance. This staggering of the primary feathers can clearly be seen in these picture of a snowy owl (Bubo scandiacus), both on the downstroke (picture to the right) and on the upstroke (picture to the left).

Snowy owl (*Bubo scandiacus*)

Rechenberg I. „Berwian" konzentriert den Wind *Sonnenenergie 2, 6-10 (1984)*
Eder, H., Fiedler, F., Neuhäuser, M. **Das Geheimnis des Storchenflugs** *BIUZ 2(56) 106 - 112. - DOI: 10.1002/biuz.201610589 (2016)*

(5) Raising the covert feathers at high angles of incidence

Photographs reveal that when birds move their wings at extremely high angles of incidence, be it during flapping flight or, more likely, during the braking phase of a landing flight, the covert feathers on the upper surface of their wings are raised (1). This may be due to the high short-term suction effect that acts on the wing's upper surface or due to a shift of the boundary layer. In such extreme cases, the layer of air that directly surrounds the surface of the wing – that is, the inner part of the boundary layer – does not move backwards across the wing, like the outer airflow, but – along the upper surface of the wing – from the back to the front. This reverse airflow forms a kind of wedge that raises the undisturbed boundary layer by pushing itself underneath: peeling off and detaching the boundary layer; leading to a collapse in the production of aerodynamic force. When the covert feathers are raised automatically, they act as a reverse-flow brake that slows down this wedge of reverse airflow as it moves further toward the front. This interpretation has been confirmed by tests using models that have been carried out in wind tunnels in Saarbrucken. Similar analyses were then conducted in Berlin for practical bionic purposes.

Black-headed gull (*Larus ridibundus*)
in summer plumage

Nachtigall W., Wedekind F., Dreher A. **Hinweise auf aerodynamische Rauhigkeitseffekte an Vogelflügelprofilen**
In: Nachtigall, W. (Hrsg.): BIONA report 3, Akad. Lit. Mainz, G. Fischer, Stuttgart: 195–218 (1985)
Patone G., Müller W., Bannasch R., Rechenberg I. **Bird flight in unsteady wind conditions. II. Technical application of the "covert-feather-effect"**
Abstracts III. Workshop Soc. for Technical Biology and Bionics, Innovationskolleg "Bewegungssysteme", Jena: 151–152 (1997)

(6) Spreading the alula

When birds fly at high angles of attack, they frequently spread their alula. By means of the resulting small slot between the alula ("bastard wing") and the bird's main wing, airflow is directed towards the wing's upper surface. This enhances the kinetic energy of flow on the upper surface. At high angles of incidence, the airflow around aerofoils is prone to collapse. Figuratively speaking, this energetic boost allows

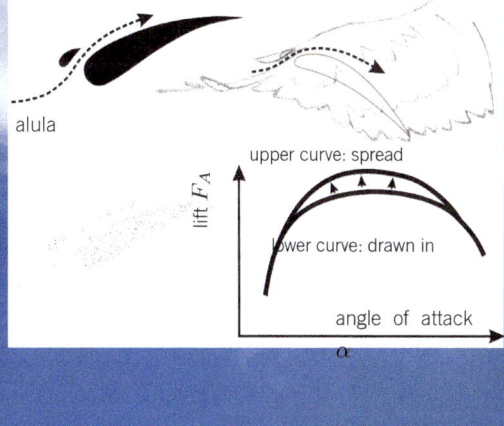

the airflow to persist and continue generating high lift. Aircraft achieve this by means of their retractable slats on the leading edge of the wings. In biology, a similar effect can be observed with the natural wings of house sparrows *Passer domesticus*, the domestic pigeon *Columba livia domestica*, the common starling *Sturnus vulgaris*, and the mallard *Anas platyrhynchos*. The additional lift generation, marked by a particularly high lift coefficient c_A, amounts to a maximum of about 15%. The characteristic curve in the sketch shows that this is only possible in a range of moderate to high levels of incidence. The alula controls a wedge-shaped section of the wing that widens towards the back, that is, at least some 10-15% of the wing surface.

Caspian gull (*Larus cachinnans*)

Nachtigall W., Kempf D. **Vergleichende Untersuchungen zur flugbiologischen Funktion der Alulaspuria (Daumenfittich) bei Vögeln. I Der Daumenfittich als Hochauftriebserzeuger** *Z. Vergl. Physiol. 71: 326-341 (1973)*

Caspian gull (*Larus cachinnans*)

(7) Optimal adjustment of a braking surface

In order to brake firmly in the air, a bird must flap its wings against the direction of the flight ("active braking") or spread out its wings and its tail to form a parachute surface with the concave side facing the flight direction. The drag coefficient of such configurations is very high, amounting to $c_w \geq 1,3$. Braking configurations of this kind – often combined with a spread alula in order to guard against the danger of sagging pressure – can often be observed during the last split seconds of a landing flight. Common examples of birds in our avifauna that regularly form such braking surfaces during landing include the common blackbird *Turdus merula* and the mallard *Anas platyrhynchos*.

Rüppell, G. **Bird flight** *Van Norstrand Reinhold Company (1975)*

(8) Spreading the tail while pulling the wings forward

If the wings are pulled forward to form a large braking surface, then the lift forces $F_{HFluegel}$ with their distance to the centre of mass $s_{Fluegel}$ produce a great, head-raising tilting torque $M_{Fluegel} = F_{HFluegel} \cdot s_{Fluegel}$.. Such a manoeuvre would cause the bird to tilt upwards along its lateral axis if it did not spread its tail downwards, which creates a tilt back downwards. That is, the tail's lift force $F_{HSchwanz}$ via the rotation distance stail produces a head-lowering tilting torque $M_{Schwanz} = F_{HSchwanz} \cdot s_{Schwanz}$ (simplified approach). If both torques are directed against each other, the bird will brake in a stable position. Such an interplay of torques has been described in studies of alpine choughs *Pyrrhocorax graculus* and common ravens *Corvus corax*.

Black-headed gull (*Larus ridibundus*)

(9) V-position of the wings

Gliding birds often hold their wings raised at an angle in a characteristic V-position. They typically adopt such a wing position during attacks or feint attacks or when they glide in close proximity to their nests. This V-position of the wings can also be observed with domestic doves *Columba livia* as they land. The domestic bird glides fast, but it sways slightly about its longitudinal axis: Its flight is relatively unstable. However, small navigational impulses are often enough to compensate for this instability during flight. This is particularly useful for attacks and precision landings.

Conclusion: Photographs of birds are not "just beautiful"

Photographs of birds can be analysed from various biological and physical perspectives. Photographs that are not easy to explain may still inspire further studies.

Photographs of bird flights are often beautiful, but never "just beautiful". They always contain informative components that are worth being examined.

Kuttner J. **Über die Flugtechnik einiger Hochgebirgsvögel** *Kosmos 43: 384–389 (1947)*
Gartmann P.W., Feder J. **Fliegen** *AT-Verlag Aarau, Stuttgart (1993)*
Ettlinger R. **On feathered Wings – Birds in Flight** *Abrams New York (2008)*

Tuck up your legs!

African oystercatcher (*Haematopus moquini*)

Tilting the distal against the proximal

These four individual shots form a series capturing the downstroke of a flying African oystercatcher, from the middle of the downstroke to the end, when the distal portion of the wing is tilted against the proximal portion. The flesh-coloured, reddish legs (*Haematopus* transla-

tes as *red-legged*) are shaded by the bird's body, which is why their colour is barely recognizable in these images. However, the photographs clearly show their position, namely stretched backwards and pulled up to the body, which minimizes drag during flight.

Not sinking further due to "ground effect"

The shearwater pictured on this page approaches the water surface with spread wings, and then glides close to the water at a distance roughly equivalent to the height of its trunk. The so-called ground effect that occurs between the bird's trunk, its pulled-down wings, and the water surface keeps the bird from sinking any further. The bird is, however, slowed by this ground effect; otherwise it could remain in perpetual motion.

Yelkouan shearwater (*Puffinus yelkouan*)

Flying with torque compensation …

Stretch those legs backwards!

Few photographs of bird flights illustrate torque compensation as clearly as those taken of flamingos with their extended necks and legs. Their centre of mass is located in the middle of the trunk, around the sternum. The bird can thus rotate about its lateral axis. If the flamingos pictured on this page did not stretch out their legs, they would rotate in a counter-clockwise direction.

"Wing salute"
Spectacular behaviour during courtship

Torque compensation avoids tilting

In order to eliminate those tilting tendencies, flamingos must move their wings in a complex manner, which would require additional metabolic energy. However, torque compensation allows them to focus their flight performance on moving forward and generating lift. Their flight muscles are relatively weak in comparison to those of other birds. Yet it is possibly due to their balanced flight that flamingos are still capable of flying distances of up to 500 km.

Only one species endemic to Europe

Despite its wide distribution, there are only six flamingo species in total, and only one of them, namely the greater flamingo (*Phoenicopterus roseus*), can be found in Europe. With a height of up to 1.60 m and a weight of 4kg, the greater flamingo is the tallest and heaviest species. They breed in large colonies, gathering in nesting areas that provide enough food and protection from predators. If the living conditions in their area worsen, they may embark on long journeys to find a new nesting site. Due to a lack of suitable nesting sites, protective measures have been taken to ensure breeding sanctuaries for flamingos, which has led to a significant increase in the flamingo population in Camargue and Cyprus.

Carotenoids colour their plumage orange-pink

Flamingos are nutrition specialists, and in the course of their phylogenetic history, their beaks have specially adapted. The beak of a flamingo can filter out and retain plankton (comparable to the sieving process of baleen whales). Flamingos also feed on small crabs. The carotenoids in the bodies of these small crabs is actually responsible for the orange-pink colouration of the flamingo's plumage.

Spectacular mass mating

It takes flamingos six years to reach full sexual maturity. Their courtship and the displays they perform to attract a mating partner arc quite remarkable. The image at the bottom on the left-hand page shows the so-called "wing salute," for which the flamingo stretches its neck and holds it still while it spreads its wings and flips up its tail. Groups of flamingos can be observed remaining motionless in this position for about ten seconds before they continue with a different form of courtship display. When this is done by several hundred birds simultaneously, the whole colony may appear from a distance to have suddenly changed its colour.

Feather parasites as evidence of phylogenetic relationship

Molecular-genetic data has shown that flamingos are related to grebes (such as the great crested grebe and the red-necked grebe). This had previously been assumed because flamingos and grebes share the same feather parasites (bird lice). It is believed that these bird lice could have been found on their common ancestors, which then evolved into different species.

Johnson A.R., Cézilly, F. **The Greater Flamingo.** *T & AD Poyser, London (2007)*
Clay T. **The Phthiraptera (insecta) parasitic on flamingos**
    (Phoenicopteridae: Aves) J. Zool 172 483–490 (1974)
http://www.flamingoatlas.org/

Cormorants are "totipalmates"

Their dense and barely pneumatized skeleton is considerably heavier than that of pelicans, to which cormorants are related. This reduces the generation of lift. Cormorants (pictured here, a European shag) are able to dive to great depths at a high speed and low energy cost.

"Jackknifing" into the water

Cormorants dive into the water like diving ducks, which are specifically heavier than dabbling ducks. Their webbed legs, which are stretched to the back as the cormorant plunges into the water, must be immersed underwater as quickly as possible, so that the cormorant can make its first paddling stroke using the momentum of diving into the water. When they are not diving and feeding the young, cormorants can often be observed sitting on poles, rocks or boughs with partly or fully spreads wings, drying their feathers. Fishers often lament the fact that these avid fish eaters – "food competitors!" – are protected by law.

European shag (*Phalacrocorax aristotelis*)

Paddling underwater

European shag (*Phalacrocorax aristotelis*)

The streamlined shape is crucial!

The cormorant (pictured here, once again, a European shag) moves underwater with a long stretched neck and an elongated, "streamlined" trunk.

Their maximum thickness is located far to the back. As has already been shown with the example of the black-headed gull, this shape ensures better flow and reduces drag. The lower the body resistance, the faster the bird can swim with a given diving power.

Bringing the legs forward

Cormorants swim underwater by merely using their paddling legs, as opposed to sulids, for instance, which also use their wings as they dive. They swing their legs forward, with the webbed toes folded up, and bring them close to the body in order to reduce counter-thrust forces that slow them down.

Only now do they start paddling

The webbed feet are then unfolded and used as paddles to swim underwater (image series to the right). The webbing between the toes is spread to its maximum extent, and they produce maximum thrust – at the highest velocity – by extending perpendicular to the body; any other position would waste energy by generating useless lateral drift.

Intuitively the best solution

This makes sense because the thrust generated is proportional to the square of the relative velocity between the webbed foot and the surrounding water, and it is at the spots where the velocity is the highest that the cormorant's paddling power is transformed into thrust without any losses (that is, without lateral drift).

Water beetles make the same optimization

Water beetles use their paddling legs for optimization based on the same principle. The legs of predaceous diving beetles (Dytiscidae) are equipped with rows of swimming hairs that enhance the generation of thrust. On each paddling stroke, the thin hairs spread out automatically due to the water pressure, and they are pushed close together again when the legs are moved forward. Whirligig beetles (Gyrinidae) have swimming setae instead of hairs. These setae spread like playing cards on each paddling stroke and are even more effective in generating thrust.

Sebaceous glands

The Mandarin duck (*Aix galericulata*) originally comes from East Asia. As with all birds, preening plays an important role. For this purpose, they are, like most bird species, equipped with a special type of oil-secreting gland, the so-called uropygial gland. These ducks may flap their wings for various reasons. They may do so to shake out their feathers after preening (which is frequently the case) or as a form of courtship behaviour. They begin to display this courtship behaviour in autumn, and they may continue to do so until spring even if they have no success in achieving copulation. The rows of covert feathers on the underside of the wing are fluffed to shake off the water. They usually lie closely on top of one another, forming a continuous profile. The cascade of free primary feathers is also nicely captured in this photograph.

On silent wings

Owls are not birds of prey

Owls – pictured here, a barn owl *Tyto alba* – are phylogenetically very old. Despite similarities in appearance, they have no closer relationship to birds of prey. Owls have very soft, loose plumage, which allows them to fly without making any sound. The upper surface of their feathers is smooth like velvet.

"Dampening" air turbulences

In addition, the leading edges of their outer flight feathers are serrated like a saw. Heavy turbulences are thus dissolved into lighter ones, and these are then largely dampened. That is why an owl's flight is practically noiseless to our human ears. Its silent flight also enables the owl to hear noises better, and it ensures that potential prey is not warned of their approach.

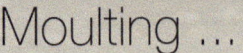

Like hair, feathers are made of the horn substance keratin
and they protect a bird's body from water and cold air. Moreover, their colours are often important for camouflage or, quite the reverse, for visual communication. Keratin provides toughness, and is adapted to a bird's flight abilities. Although feathers are light, their total weight may be twice as much as the remaining weight of a bird's body. Feathers also wear off over time even though they are continuously preened, oiled, and freed from parasites.

Wearing off requires occasional shedding ("moulting")
The growth of new feathers is always hormonally regulated. In addition to that, many young birds must moult to reach adult plumage, and there are also species that shed their regular plumage to grow a nuptial plumage before the breeding season. The hormonal regulation is often subject to the cycle of the seasons. Regardless of the type of moulting, the capacity for flight should be retained at any point even if the wing's large flight feathers are shed and change

in the process. Many water birds like ducks, flamingos, grebes, and cranes shed all their flight feathers at once and thus lose their capacity to fly for several weeks. They typically hide during that period or withdraw to places that are not accessible to their enemies.

Moult gaps have a detrimental effect on flight
Corvids and cormorants have been observed to fly even with large moult gaps. However, experienced observers will notice a decline in their ability for fine control during flight; these birds will fly less smoothly and can stay in the air for shorter durations. There have been experiments in which individual flight feathers are tied together in order to create artificial gaps. The flight of the birds used for the experiments resembled that of birds that are actually moulting. The change in flow conditions presents a considerably greater demand on the bird's energy. The same applies to plumage that is dripping wet (image below).

... and theories on the evolution of feathers

189

Feathers did not evolve from the scales of dinosaurs.

In the past it was assumed that scales modified into feathers in the course of the evolution of birds, but a more recent hypothesis claims that they evolved independently alongside scales. This hypothesis has been obtained from numerous findings of fossilized theropods (dinosaurs), which belong to the stem group from which modern birds evolved. During the initial stages of the evolution of feathers, we find simple filaments. Later, different lineages of theropods evolved different types of feathers. Some of these feathers resembled the fluffy down of modern birds, some had symmetrically arranged barbs. Other species of theropods grew long, stiff ribbon-like feathers or broad filaments that bear no resemblance to the plumage of birds today.

Genetic evidence against the "scale theory"

A genetic study of feather growth among modern young birds has shown that feathers cannot have evolved from scales. The young bird's feather placode, as the formative tissue is known, may resemble the formative tissue of scale-bearing reptiles, but only one single mutation developed embryonic rudiments of feathers that grew vertically through the skin rather than parallel or inclined to the skin's surface. The first filaments only required minor modifications in order to evolve into increasingly complex feathers. There have also been findings of fossilized dinosaurs (such as the famous *Tianyulong confuciusi* from the family Heterodontosauridae), which already sported bristle-like structures growing between their scales on the back.

Why has evolution brought about feathers?

The functions that feathers perform in modern birds do not provide an explanation as to their origin. The thesis relating the development of feathers to improved thermoregulation is probably the most plausible one given that many dinosaurs were homeothermic, maintaining a stable internal body temperature. So the thermal protection provided by feathers must have been quite helpful. According to a more recent theory, feathers also serve as a method of waste disposal (detoxification) in a bird's metabolism. Birds ingest sulphurous amino acids via their food, which they cannot completely dispose of through their digestion. However, these amino acids can bind to and be removed via other substances – such as the keratin of feathers. So like the scales of reptiles, feathers are believed to have originally functioned as a kind of rubbish bag and only later they began to serve as means of thermal isolation and transportation.

Stresemann, E. und V. **Die Mauser der Vögel** *Journal für Ornithologie 107, Sonderheft. Berlin (1966)*
Reichholf, J.H. **Der Ursprung der Schönheit. Die biologischen Grundlagen des Ästhetischen.** *Verlag Beck, München (2009)*
Zheng, Xiao-Ting; You, Hai-Lu; Xu, Xing; Dong, Zhi-Ming **An Early Cretaceous heterodontosaurid dinosaur with filamentous integumentary structures** *Nature 458 (7236): 333–336 (2009)*
http://www.nationalgeographic.de/reportagen/ein-wunder-der-evolution-wie-die-natur-die-feder-erfand

"Eagle eyes"

The physical maximum

The resolution capacity of an eagle's retina is up to 4 times greater than that of a human. Eagles are thus capable of spotting a mouse from a height or distance of over 3 km. Moreover, in relation to their body size, the eyes of eagles are 1.4 times bigger than that of other birds of the same size. Their visual acuity is unbeatable within the whole animal kingdom. In addition, an eagle eye has two foveae, which are the points of sharpest vision on the retina (one of which is directed to the side and the other forward). So, like humans, raptors can look forward while their visual field and spatial perception is significantly enhanced by their additional fovea looking to the side. With 4 to 5 types of cones in their eyes (the human eye has only 3), eagles can perceive colours better. At night, however, birds of prey cannot see particularly well.

Owl eyes

In order for owls to be able to hunt for prey under low light conditions, their eyes must be extremely light-sensitive and large: They occupy about one third of an owl's head. If human beings had eyes of similar dimensions, they would have to be the size of an apple. The eyes of owls are, in fact, about 2.2 times larger than those of birds of roughly the same size. Their visual fields overlap by up to 70 %, which provides owl with very high spatial resolution.

Behind their retina, owls have a light-reflecting layer that makes their eyes glow at night when they are hit by light. However, not even owls can see under conditions of complete darkness. In bright light, on the other hand, the pupils of owls contract into very small circles in order to prevent damage to the eye.

Bald eagle (*Haliaeetus leucocephalus*)

Snowy owl (*Bubo scandiacus*)

The masters of hovering flight

Depicted by the ancient Nazcas
Hummingbirds are typical birds endemic to the Americas. Even the mysterious Nazca people immortalized the tiny hovering bird in their geoglyphs in the Peruvian desert. The inner tail feathers of the sub-family Phaethornitinae (pictured here, a *Phaethornis longirostris*) are extremely elongated.

Shimmering colours

The iridescent feathers of hummingbirds bear several layers of microscopically thin horny lamellae. If these feathers are hit by light from a certain angle of attack, the lamellae will reflect the light to produce the hummingbird's splendid shimmering colours.

Hill, G.E. & K.J. McGraw (ed.) **Bird Coloration: Mechanisms and measurements**
Harvard Univ. Press (2006)

Manoeuvres at 50 wingbeats per second

Analysis of a wingbeat

The female black-chinned hummingbird feeds from a big-sage flower *Lantana camara* in hovering flight. The photographs on this page capture four typical wing positions during different periods of the wingbeat cycle. The image to the top right shows the beginning of a downstroke. The wings are still held up high, but the left-hand wing is already turning into position for the downstroke. The image to the top left shows the middle of the downstroke; the left wing appears to be paper-thin, because it was photographed with the narrow edge facing the camera. The image to the bottom left is taken at the lower reversal point as the wings transition into the beginning of the upstroke; the distal wing area is "dragged behind". This whole process is repeated about 50 times a second.

Study of the flight of a female Rufous hummingbird. These photographs were taken in more "northern regions", on Vancouver Island. Hummingbirds fly thousands of kilometres every year to reach their winter habitats in Mexico. In order to save energy, they enter into a state of torpor at night.

Skilled in flying and diving

Beak serves to "clamp fish"

The Atlantic puffin is only slightly larger than a common dove. It breeds in burrows and on cliffs of the northern Atlantic (here: Grimsey Island to the north of Iceland). Its body shape produces particularly little drag. The peculiarly shaped triangular beak is extremely high but very narrow. It has a series of grooves which allow the Atlantic puffin to carry several fish with its beak – clamping them one after another (see image above).

Flyers with stamina

The winter habitats of the birds pictured on this page are located in Newfoundland, which is some 2500 km from Grimsey Island. The trunk of the Atlantic puffin appears to be disproportionally chubby. Its maximum thickness occurs further back, which minimizes the drag generated by the bird's trunk. When the Atlantic puffin takes off from the surface of the water, the bird, which must cope with high wing loading, moves its wings with as much force and amplitude as possible. It sometimes pushes itself off the surface with its legs.

Special technique

Wing twisting: The leading edge is increasingly pulled downward from the base to the tip. This form of twisting is also called "pronation". It is supposed to avoid a collapse of airflow on the outer surface.

Take-off jump – always exhausting

In order to gather speed, the grey heron *Ardea cinerea* (above) makes a powerful take-off jump by bending and then stretching its legs. The first downstroke is done with as much amplitude as possible. The third to last image clearly shows the necessary wing twisting towards the tip; in the second to last image, the distal portions of the two wings almost touch. The wings cannot be pulled further down. This may be the most exhausting kind of wingbeat, only surpassed by stationary hovering flight.

Vertical take-off

The great egret *Ardea alba* (below) takes to flight from the water almost vertically with only one powerful wingbeat. The wings function like a horizontal propeller in this case.

Both image series: 30 frames per second

Long legs

Necessary but often obstructive

Evolution rarely proceeds in a linear fashion when adapting certain structures to their functional requirements. Herons have very long legs because they hunt for prey at the water's edge. The length of their legs determines how deep they can move into the water. Since the various heron species differ in the length of their legs, they can divide their hunting grounds. Many herons build their nests in high trees. Long legs can be very obstructive when moving through the branches of a tree, and so selection had to strike a balance between long legs for fishing and short legs for accessing the nest and breeding. It is quite impossible to be a flight acrobat with such "landing gear".

Black-headed heron (*Ardea melanocephala*)
Pied crow (*Corvus albus*)

Attacking at a consistently high velocity

The osprey *Pandion haliaetus*, as opposed to the golden eagle, for instance, has barbs on its feet, with which it can catch and hold slippery prey like fish. As the osprey flies over a fish, it rapidly kicks its legs backward while its trunk moves forward at the same velocity, and the legs are lowered in the process. From the perspective of the fish, these birds grab their prey very effectively from directly above. The series of images on this page shows how the osprey approaches the surface of the water at a constant velocity of about 25 m/s (90 km/h) and then "rams" the fish. The image series to the right shows the grasping of the prey on the water surface. The osprey first dives underwater and then rises again with very powerful, flailing flaps of its wings. As it flies away, it turns the fish in its feet so that its head faces forward in order to minimize the drag produced by the prey.

Not always entirely safe

Sometimes the fish is too big but the osprey cannot let go of its prey due to its deeply inserted claws. The skeleton of an osprey attached to the body of a large carp testifies to a tragedy of this kind.

Astonishing sense of balance

Not only small singing birds, but also large birds are masters of balance. The osprey pictured on this page has landed on top of a weather station. The bird balances itself against the wind, trying to move its body's centre of gravity vertical above the point of touchdown and to hold it there by twisting its trunk.

Run-up to take a nosedive

Quarter turn in the last tenth of a second
Below: Details of a "dive" into the water: The pelican rotates at the very last moment (like the bee-eater on p. 28). Its velocity is considerably lower than that of the nosedive pictured to the left.

The acceleration of gravity is not enough
It usually takes no more than 1.5 s to cover a distance of 10 m falling freely from a starting velocity of zero. This pelican, however, travels the same distance in only half the time (this composite photograph consists of 9 images taken at an interval of 1/12 of a second). This is due to the fact that the pelican uses its flight velocity to launch into the nosedive. The bird pictured here flies initially at a speed of 8 m/s and then accelerates to 15 m/s (about 55 km/h).

The search for food and competition
The picture at the top right shows the "launch phase", when the pelican transitions into the nosedive. The images below show two pelicans engaged in a competitive battle. No extreme nosedives can be seen in the images on this page.

A successful species

Smooth landing ...

Image to the top right: What appears to be so simple is actually revealed by a close analysis to be based on calculated behaviour and a mastery of aerodynamics.

The final downstroke before the touch-down of the legs is a powerful braking flap with fully spread wings, widely spread tail, and splayed alula. The splayed feathers on the distal portion of the right wing can clearly be seen in the third to last image of the final upstroke. Each of these feathers contributes a lift component and thus prevents a sudden drop to the ground.

... and suddenly quite impetuous

Image to the bottom right: Doves are generally considered to be peaceful animals, but they can become quite impetuous, especially during mating and territorial fights. In both cases, they make intensified use of their wings.

7 **Bats**

Flying mammals

It was only about 50 million years ago that bats and flying foxes took to the skies. Insects have been flying for a period seven times as long, and birds for a period four times as long. The pterosaurs, which had been flying through the air for some 170 million years, disappeared due to drastic climate changes towards the end of the Mesozoic Era. Navigation by means of ultrasound has enabled bats to conquer darkness.

Evolution of bats …

The only active flying mammals

Bats (Chiroptera) are the only order of mammals which have evolved the capacity for active flight. They are traditionally divided into two suborders: microbats (Microchiroptera) and megabats, or flying foxes (Megachiroptera). While microbats can be found around the world, flying boxes are only endemic to sub-Saharan Africa, in the Nile Valley northward to Egypt, and from South-East Asia to Northern Australia. One species from northern Egypt can even be found near the eastern Mediterranean (e.g. Cyprus).

Bats use ultrasound to navigate

Bats are truly nocturnal flyers and only they navigate by means of ultrasound, which can reach a frequency ranging from 10-140, rarely even up to 200 kHz. Their calls usually consist of a series of five or more different sounds, which may last from about a second to just one hundredth of a second. Adult humans can typically detect sound frequencies lying 16 Hz and 18 kHz, and so we can only barely hear the calls at the lower end of the scale as chirping noise. Bat detectors can make the ultrasonic calls of bats audible to the human ear. They transform these calls into sound waves with a lower frequency that falls within the range of human hearing. Flying foxes do not emit ultrasonic calls. Only the Egyptian fruit bat produces clicking sounds to achieve some spatial orientation in caves. Flying foxes are active at twilight and rely on their visual senses for navigation. They have relatively large eyes for this purpose, which are noticeably larger than those of microbats. Flying foxes eat fruits and lick nectar from flowers, whereas microbats primarily feed on insects. However, there are also large species that hunt frogs, fish, and small mammals or even suck blood (*Desmodus*).

Bats have unique wings

The family of leaf-nosed bats has evolved several genera endemic to the tropical regions of South and Central America that secondarily feed on fruits and the nectar of flowers (Phyllostomidae, Glosso-phaginae). The morphological structure of their wings is entirely unique. They have a flying membrane that spans between their highly elongated 2nd to 5th fingers and then extends to their hind legs. In microbats, this membrane runs even further down to their elongated tail, where it forms a small tail membrane. Flying foxes have a shortened tail, sporting only a very tiny tail membrane at best. Microbats have only one claw extending from their thumb (= 1st digit), whereas the larger flying foxes have another claw on their second digit. These claws allow bats to cling onto surfaces and to stabilize themselves while walking. Bats always spot their prey during flight and catch them with the help of their flying membranes.

Fossilized bats

In order to find out when bats emerged in the past and when and how they began to fly, we have to rely on fossil discoveries. Bats as active flying mammals are only rarely preserved as fossils. An exception is the Messel pit near Darmstadt in Hessen with numerous fossil discoveries from the middle Eocene. 700 species from four

different families have been found at this natural World Heritage site. The exceptional preservation of their stomach content, inner ears, and the flight apparatus allowed conclusions to be drawn about the life of these animals during the Eocene. The bats that are considered to be the most primeval based on the morphology of their teeth and bodies belong to the family Archaeonycterididae, which includes the genus *Archaeonycteris*: Like the *Archaeonycteris trigonodon*, for instance, they had a short, straight lower arm, and they sported an additional claw on their index finger. The about 700 species found up to date resemble modern bats in the wide range of variation of different parameters relevant to the aerodynamics, such as the shape of the wings and wing loading.

… and of their flight capacity

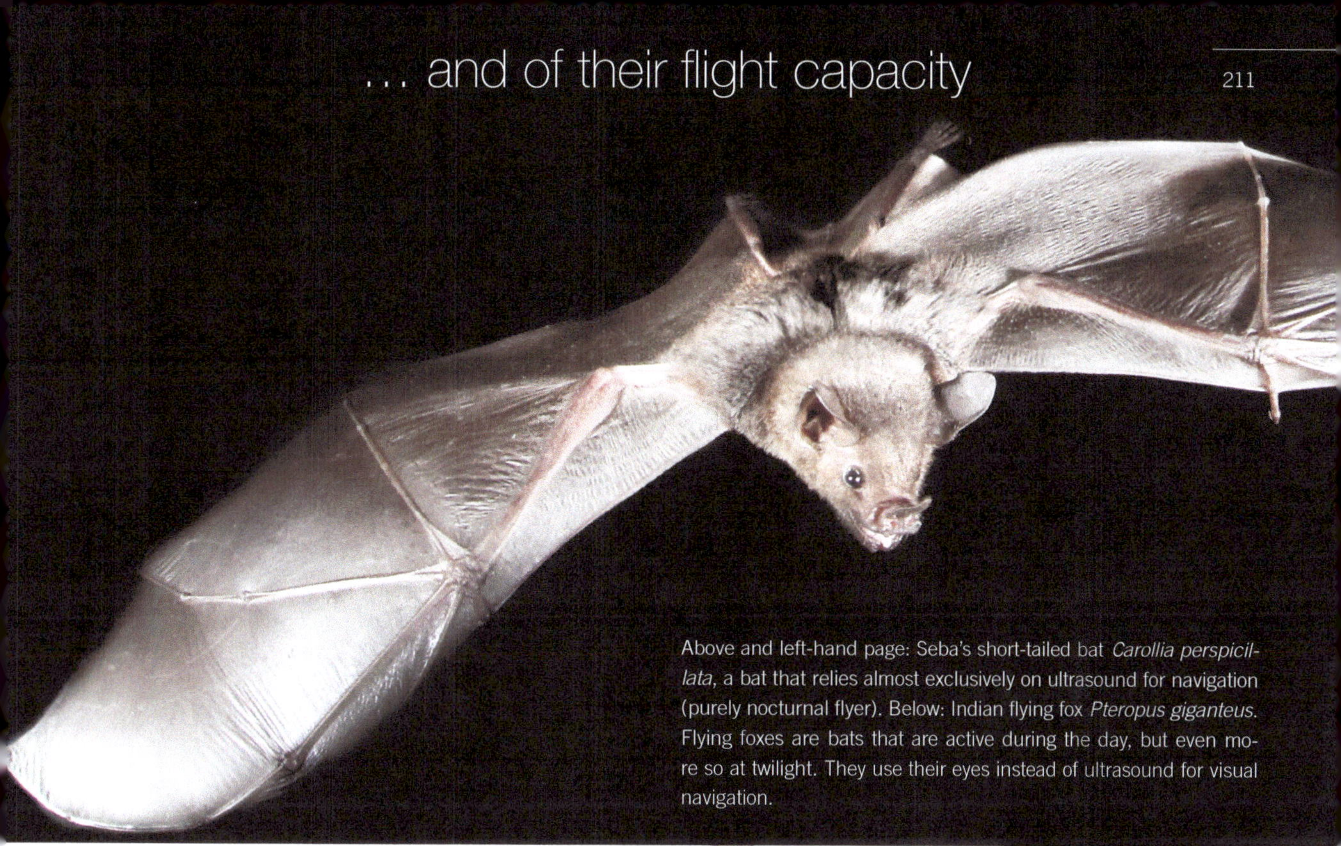

Above and left-hand page: Seba's short-tailed bat *Carollia perspicillata*, a bat that relies almost exclusively on ultrasound for navigation (purely nocturnal flyer). Below: Indian flying fox *Pteropus giganteus*. Flying foxes are bats that are active during the day, but even more so at twilight. They use their eyes instead of ultrasound for visual navigation.

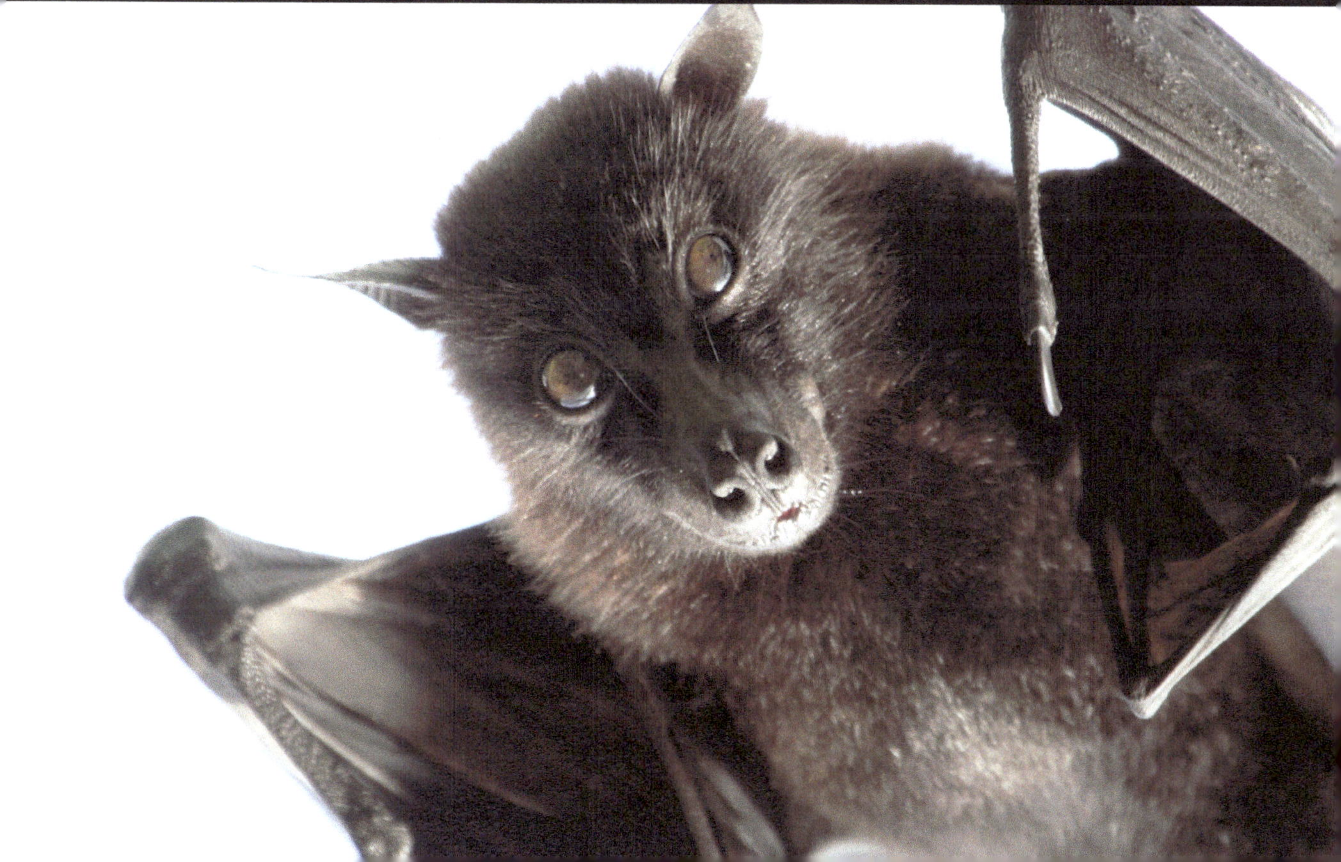

Primeval bat and "more modern versions"

A more than 52 million year-old primeval bat

In 2003, the fossil of the previously unknown bat species *Onychonycteris finneyi* was discovered in Wyoming. What is so remarkable about this fossil discovery is the very short wing skeleton of this bat. That is to say, its hand is significantly smaller than that of all known species of microbats, flying foxes and - most strikingly - all paleontological discoveries found to date. This fossil, therefore, represents a precursor of a typical modern bat.

The oldest "modern" bat

Paleo-biologists from the Senckenberg research institute have examined the fossilized remains of bats found in the Messel pit near Darmstadt for years. The discoveries may have been numerous and spectacular, but the fossils from Germany are some 5 million years too young to qualify as a primeval form of the bat.

The oldest fossil discovery of flying foxes only dates from the Oligocene (some 34-23 mya) in Italy (*Archaeopteroptus transiens*), and there was another discovery from the Miocene (some 23-5 mya) in Africa.

Seba's short-tailed bat (*Carollia perspicillata*)

Moths had to adapt

The evolution of bats had a considerable impact on the further development of the insect world. All nocturnal insects were suddenly confronted with powerful predators which could easily detect their prey with their ultrasonic calls. As an "evolutionary reaction" to these new circumstances, almost all moth species developed special ears, so-called eardrum or tympanal organs, which enabled them to hear the calls of bats. The bizarre wing shapes of some moth species may also have evolved to make it more difficult for bats to detect them.

Wings or ultrasound first?

People have pondered upon this question for a long time. In order to arrive at an answer, the few fossil remains that have been found must be examined closely.

The cochlea must be adapted to ultrasound

The fossilized bats from the early Tertiary have shown that these animals were already capable of flying, because their wings barely differ from those of more recent species. The questions as to whether they could already navigate with ultrasound could be clarified by measuring the cochlea of their inner ear. The cochlea is well preserved in the fossil remains of these bats. Bats with ultrasound recognition have a cochlea with a characteristic shape that clearly differs from the cochlea of flying foxes, which cannot detect ultrasound.

Fossils provide insight

Due to the good preservation state of Eocene bat fossils, it has been possible to create a detailed reconstruction of their inner ears using X-ray analysis. The result: Eocene bats (such as *Onychonycteris*) could only use low-frequency echolocation for spatial orientation, which was not enough for the hunting of flying prey.

Has ultrasound navigation evolved several times?

Molecular-genetic analysis has proven that the group of horseshoe bats is, quite surprisingly, more closely related to flying foxes than it is to the other bats. That is to say, they share a last common ancestor with flying foxes. From these results, it follows that either flying foxes lost the ability to navigate with ultrasound or horseshoe bats and the other bat species developed this ability independently of each other.

Ultrasound: horseshoe bats vs. other bats

Horseshoe bats call at a much higher ultrasound frequency, and their calls are very long (several milliseconds) and stay at a constant frequency. The other bats produce ultrasonic calls that are mostly at lower frequency and act as frequency-modulated location signals (shorter range, but better information on flying insects). Horseshoe bats emit their calls through their nose: They have wrinkles and folds around the nostrils that are shaped like a horseshoe. These folds serve to direct and focus their calls. The other bats emit their calls through their mouths. All this supports the theory that horseshoe bats have developed their ability to navigate with ultrasound independently from other bats.

How did the wings evolve?

Of course, the question remains as to how the wings of bats may have evolved, since no fossilized transitional remains have been found so far as evidence of any intermediary stages. Developmental-genetic research has at least shed light on those mechanisms that enabled the evolution of a mammal's front limbs into wings: Both the development of flying membranes and the functional modification of the finger bones can be triggered by relatively simple changes in the regulation of gene expression. The subfeatures involved in these changes, such as the muscles, the nerves, and the blood vessels, can automatically grow along with the changes and reorganize themselves, so extensive morphological changes that enabled gliding flight could have occurred with relative ease and over a short period of time. The lack of fossilized transitional forms is thus less surprising.

Simmons N.B., Seymour K.L., Habersetzer, J., Gunnel G.F. **Primitive Early Eocene bat from Wyoming and the evolution of flight and echolocation** *Nature 451, 818-821 (2008)*

Dietz C., von Helversen O., Nill, D. **Handbuch der Fledermäuse Europas und Nordwestafrikas** *Kosmos Naturführer (2007)*

Jones G., Teeling E.C., Rossiter S.J. **From the ultrasonic to the infrared: molecular evolution and the sensory biology of bats** *Frontiers in Physiology May 2013 Volume4 Article117: 1-16 (2013)*

Smith T., Rana R.S., Missiaen P., Rose K.D., Sahni A., Singh H., Singh L. **High bat (Chiroptera) diversity in the Early Eocene of India** *Naturwissenschaften 94 (12): 1003–1009 (2007)*

Above a "flat-nosed" common pipistrelle (*Pipistrellus pipistrellus*)
Below a greater horseshoe bat (*Rhinolophus ferrumequinum*)

Seeing with the ears

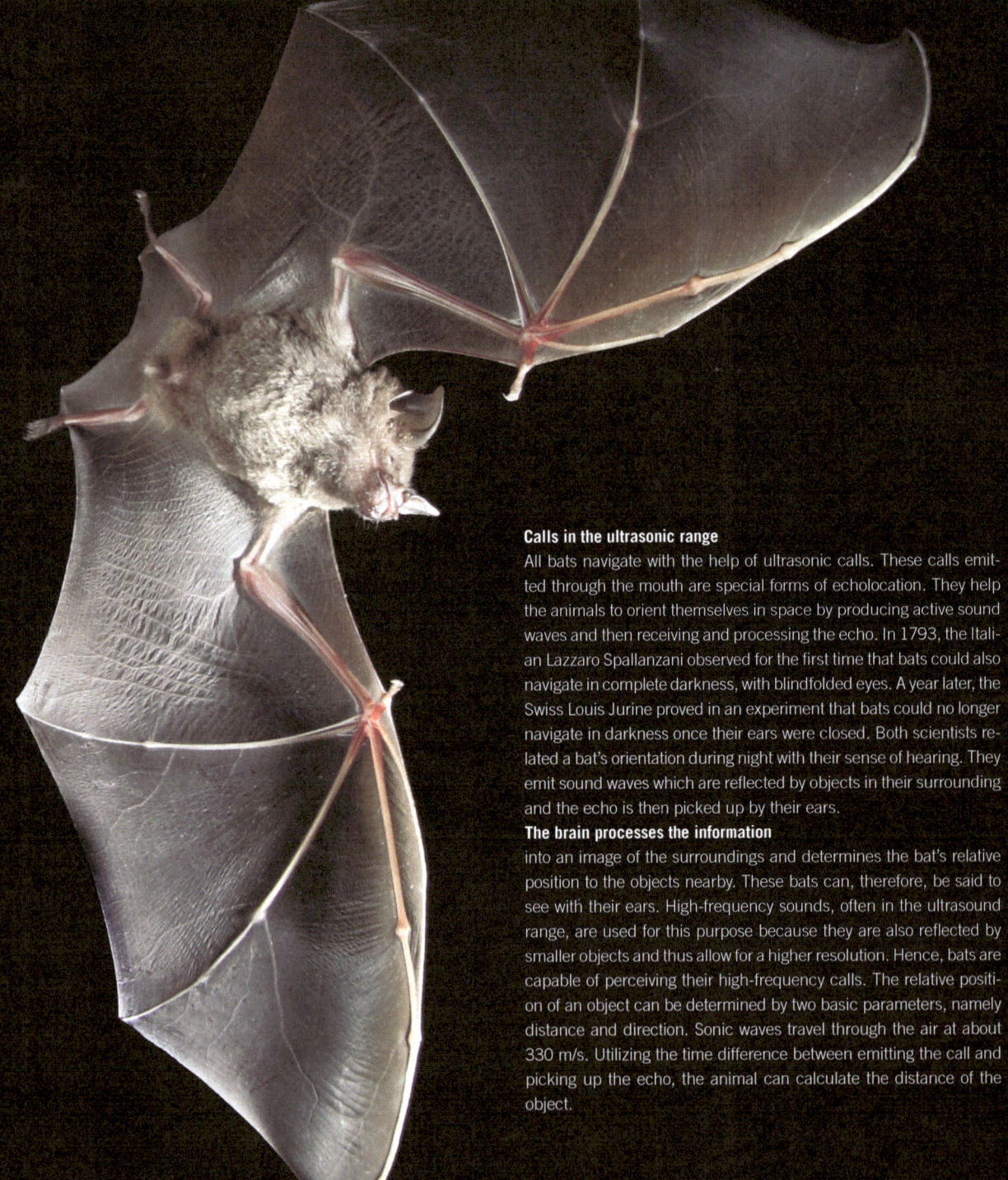

Calls in the ultrasonic range

All bats navigate with the help of ultrasonic calls. These calls emitted through the mouth are special forms of echolocation. They help the animals to orient themselves in space by producing active sound waves and then receiving and processing the echo. In 1793, the Italian Lazzaro Spallanzani observed for the first time that bats could also navigate in complete darkness, with blindfolded eyes. A year later, the Swiss Louis Jurine proved in an experiment that bats could no longer navigate in darkness once their ears were closed. Both scientists related a bat's orientation during night with their sense of hearing. They emit sound waves which are reflected by objects in their surrounding and the echo is then picked up by their ears.

The brain processes the information

into an image of the surroundings and determines the bat's relative position to the objects nearby. These bats can, therefore, be said to see with their ears. High-frequency sounds, often in the ultrasound range, are used for this purpose because they are also reflected by smaller objects and thus allow for a higher resolution. Hence, bats are capable of perceiving their high-frequency calls. The relative position of an object can be determined by two basic parameters, namely distance and direction. Sonic waves travel through the air at about 330 m/s. Utilizing the time difference between emitting the call and picking up the echo, the animal can calculate the distance of the object.

have specialized in hunting fish (*Noctilio*), species that suck blood (*Desmodus*), and species feeding on fruits or the nectar of flowers (Glossophaginae).

Evolutionary adaptation of plants and insects

The precise spatial orientation and object recognition of bats is quite astonishing to us human beings. The family of leaf-nosed bats also includes species that hover in front of flowers while feeding like hummingbirds. They detect their flowers not only by their particular scent but also by using echolocation. For this purpose, some flowers have developed special ultrasound reflectors, which make it easier for bats to detect them. Some insects that are frequently hunted by bats have evolved adaptations that allow them to detect the ultrasonic calls of bats to let themselves drop to the ground quickly and escape their predators. These include many moth species that have developed special tympanal organs acting as "ears" to hear the calls of bats.

The frequency of most sonic waves is too high for us

Bats produce echolocation sounds in their larynx, at a frequency ranging from 8-160 kHz depending on the species or genus, though most of these calls are beyond our hearing capacity. These sounds leave the bat's body through the mouth, or through the nose in the case of the horseshoe bats (Rhinolophidae). Horseshoe bats have special folds at their nose which focus the sonic waves. To receive the echo, bats are equipped with highly developed ears and large pinnae, which frequently sport a broad, movable tragus. The vertical position of the object relative to the bat is determined either by means of interference patterns created by the tragus or by independently raising and lowering their pinnae. Bats adapt their echolocation calls to the prey's distance. In order to detect prey at a long distance, they emit long, narrow-band calls of only a few frequencies. For short-distance location, on the other hand, they emit broad-band calls that span many frequencies and last less than 5 milliseconds. These calls allow them to determine the exact location of an object. Such bats are also known as FM (frequency modulated) bats. There are also bats that only produce echolocation calls with constant frequencies. These are classified as CF (constant frequency) bats.

Leaf-nosed bats – from the "vampire" to the fruit eater

The bats pictured on this page belong to the family of leaf-nosed bats (Phyllostomidae), which are only found in the south of the USA and in Central and South America. They are a diverse family of many species, ranging from bats feeding on insects to astonishingly large predators hunting for small mammals, such as the spectral bat *Vampyrum spectrum* with a wing span of 1 m. This species is the largest bat. Only some species of flying foxes endemic to Africa and South Asia, such as the Indian flying fox with a wingspan of almost 170 cm, can grow to become larger than that. In addition to those leaf-nosed bats feeding on insects or small mammals, there are also species that

Analysis of a bat's hovering flight

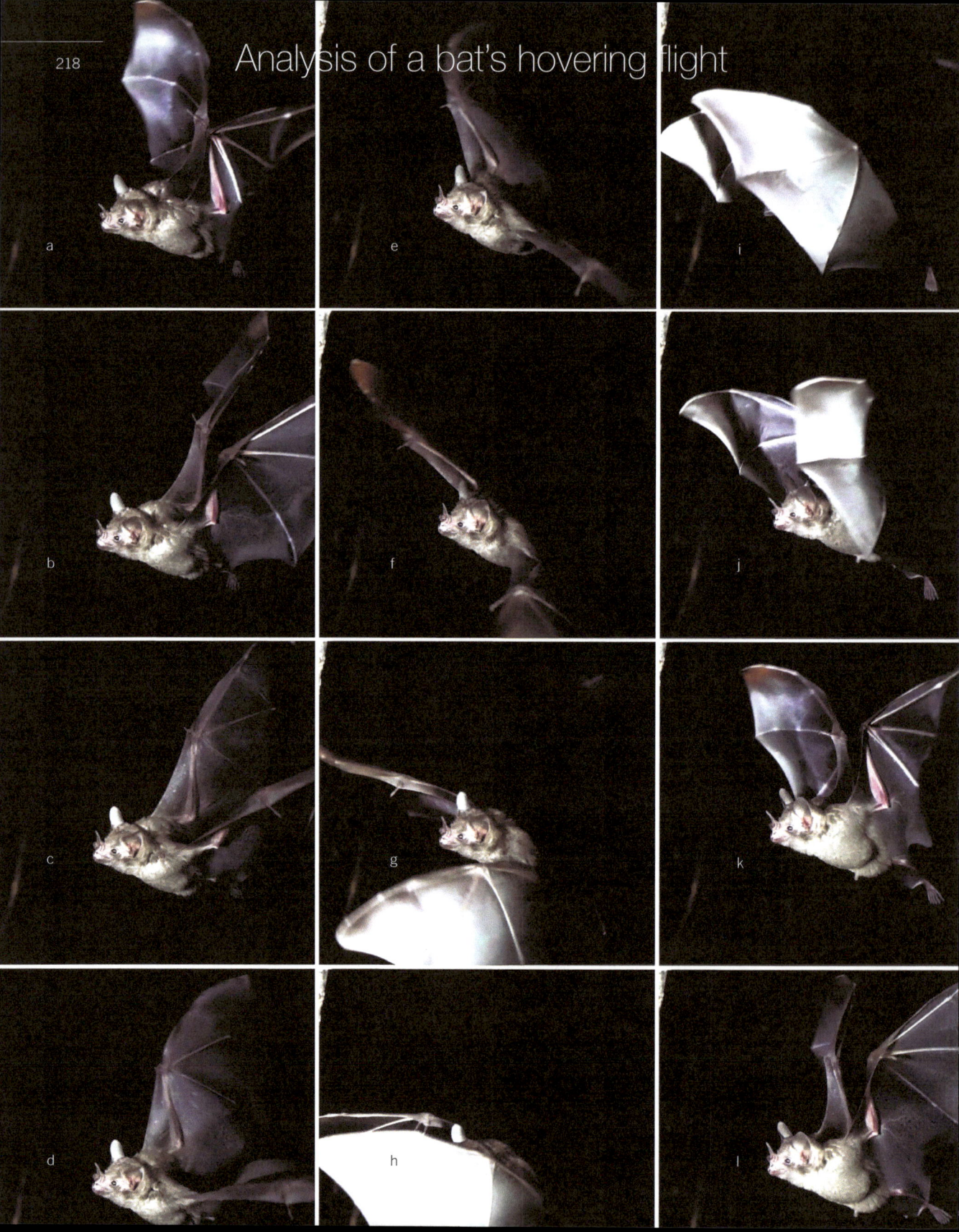

Both image series show a complete wingbeat cycle (1/11 second)

of a leaf-nosed bat *Carollia perspicillata* flying freely in hovering flight. The images were successively shot from the side (left-hand side) and from the front (right-hand side). The wings beat from the top back to the bottom front, and then move back in a different configuration. The image labels refer to the left-hand series, but to a large extent they can also be applied to the image series to the right. At 120 frames per second, a wingbeat lasts $11/120 \approx 1/11$ s. A small bird hovering in front of a feeding dish reaches a wingbeat frequency that is about twice as high; the wingbeat frequency of small hummingbirds with their robust wings is even five times higher. A bat's delicate flying membrane with its fairly low wing loading makes up for the low wingbeat frequency by having a large surface.

From the upper reversal point into the downstroke

Images a and b show the beginning of the upper reversal phase, and image I shows the end. The wing is here brought into a downstroke position, namely in a way that enables the wing to generate good lift forces right from the beginning of the downstroke. The downstroke is captured by images c to h. The upstroke begins with image i and ends approximately with image l. That is, the upstroke is slightly shorter than the downstroke. Moreover, the movement of the wings on the upstroke clearly differs from their movement on the downstroke.

The downstroke is easy to understand

As one would expect, the airflow during the downstroke works against the morphological underside of the wing. Pressure that is exerted on the wing from below causes the membranes between the fingers to bulge, first only slightly, but soon to their full extension, and in this configuration, the wings nearly hit each other at the bottom front – but only nearly so, they do not actually touch each other. The highest lift is produced at the middle of each stroke, when the wings extend vertically from the bat's trunk. They are slightly curved to the top towards the tip, which are most strongly affected by the aerodynamic forces.

The airflow during the upstroke is very different

because it works against the morphological upperside of the wing. The portions of the wings that are closer to the body first move at an angle to the top; the outer segments of the wing increasingly fold against the inner segments. This may look quite strange. At this point, the wing also generates lift due to its angled position. However, the "gear technology" is not as ideal as during the downstroke. This possibly explains why at some point during the upstroke, the wing relies on its tip for enhanced lift generation.

"Like a cracking whip"

The outer third of the wing, which is dragged behind relative to the remaining portion of the wing, provides enhanced lift generation. If it is positioned at an angle, it can move very quickly and since lift increases to the square of flow velocity, this configuration is very effective. It calls to mind the role of the distal portion of a bird's wing with its spread-out cascade of free primary feathers, which are struck by the airflow at an angle. This can clearly be seen when observing seagulls in slow flight (see page 56f.). They also exhibit this "whip-like unwinding." Evolution has thus produced a mechanism that seems universally applicable.

M. Prigg **Watch a fruit bat fly and a hummingbird hover:**
Hypnotic animations reveal exactly how different animals take to the skies
http://www.dailymail.co.uk/sciencetech/article-2776970 (2016)

Flying foxes have good eyes

Flying foxes are mostly active at twilight, but they usually start their flights in search for food when there is still daylight. As opposed to purely nocturnal bats, they have large eyes, with which they orient themselves visually.

Protection against the rain

Above: Short-nosed Indian fruit bats from South-East Asia *Cynopterus sphinx*, which only reach a size of about 10 cm, typically gather into small groups for sleeping. By biting the midribs of leaves and thus causing them to collapse, they form enclosed tent-like spaces under which they can roost shelteredagainst the rain.

Living in large groups

Below: Large flying foxes (pictured here a species from the Pteropodini group) are sometimes also called fruit bats or Old World fruit bats. They spend most of the day roosting in groups on high trees, from where they take to flight to search for food.

Indian flying fox (*Pteropus giganteus*)

Very straightforward during mating

In large colonies, males tend to be very straightforward when it comes to selecting a female mating partner and copulation (see image above and 93).

Toe claw with double function

The claws on their toes are adapted for grasping branches the size of a small finger. During flight, the clawed thumbs are spread away rather than folded in. This makes the boundary layer more turbulent and thus easier to detach.

Neuweiler, G. **The Biology of Bats** *Oxford: Oxford University Press (2000)*

The only "barely European" flying fox

The Egyptian fruit bat *Rousettus aegyptiacus* is native to Egypt and the Arab peninsula, but it can also be found in Cyprus. With a body length of 15-17cm, it is only of moderate size compared to other flying fox species, but it is still larger than all European bats.

Click sonar similar to that of bats and dolphins

The Egyptian fruit bat has become famous for clicking its tongue inside caves as a form of echolocation. This active technique of echolocation ("click sonar") emits sonic waves by means of subtle tongue-clicks and then processes the reflected echoes. Comparisons may be drawn to the echolocation of bats and dolphins, but these two animals use their vocal chords instead of their tongues to produce these clicking sounds. This type of spatial orientation can also be observed in blind people.

Right-hand side below: A young bat has accidentally flown to the ground and takes some time now to get back into a hanging position, which allows these animals to take to flight quickly and effortlessly. The photograph clearly shows the flying membrane between its hind legs, which is used for precise navigation during flight.

Richard A. Holland, Dean A. Waters, Jeremy M. V. Rayner **Echolocation signal structure in the Megachiropteran bat Rousettus aegyptiacus Geoffroy** *Journal of Experimental Biology* *2004 207: 4361-4369; doi: 10.1242/jeb.01288 (1810)*

8 The fascination remains

The topic is and will continue to be exciting

The fact that animals – and ultimately humans as well – can rise to the air as if gravity did not exist has always exerted enormous fascination on us. Biophysics has allowed us to understand and analyse this fact. Moreover, it is quite "typical" for the stunning capacity of evolution to use every imaginable niche and push it to its limits by placing, step by step, "one stone upon another."

From the cradle to the grave

Distributing eggs as effectively as possible
This female of the tropical swallowtail flies to the host plants of its larvae in order to lay its eggs, and it will lay only one or two eggs at a time. Many females first examine the leaf to see if there are already eggs by other females in order to avoid laying too many eggs on one leaf. The eggs are thus distributed over a wider area, which lowers the risk of all the larvae hatching from the eggs being eaten by predators or becoming infested with parasites.

A small crab spider ...
can spell death. The camouflaged spider hides on flowers and leaves, waiting for a chance to ambush and "kill" its prey by grabbing it in a suitable moment and biting it. Their chelicerae are highly venomous (see also 228).

Long proboscis vs. long nectar spurs

Extremely long proboscis, which keeps the hawk moth at a distance from the flower

The Convolvulus hawk moth *Agrius convolvuli* from the Sphingidae family of moths is one of the largest moths of our regions, and like all hawk moths, it feeds on flowers in hovering flight. In order to reach the nectar which lies very deep inside the flower, its proboscis can reach a length of 10cm, up to a maximum of 14 cm, which almost equals its wingspan. The animals thus do not have to get too close to the flower in order to feed on it: There could be dangerous spiders hiding there.

Nectar spurs became longer in order to draw the animals back closer

However, the successful pollination of many flowers (e.g. tropical orchids) requires animals to fly closer. As a result, flower developed increasingly long nectar spurs in order to force hawk moths to move closer to the flower. This, in turn, caused the hawk moth's proboscis to grow even longer. The species *Xanthopan morganii*, which can be found in Africa and Madagascar and is commonly called Morgan's sphinx moth, is famous for having a proboscis that can reach a length of up to 26 cm. The *Xanthopan morganii* hawk moth was not yet known during Darwin's time, but from his observations and experiments with the Madagascan orchid *Angraecum sesquipetale*, which has a nectar spur of up to 28 cm length, the natural scientist surmised that there must be a moth with a proboscis long enough to reach the nectar and pollinate the flower.

Extensive migrations at high travel speeds

The Convolvulus hawk moth, which can be found in our regions, is actually native to the southern Mediterranean, the tropical regions of Africa, and Asia. However, every year, this hawk moth species undergoes extensive migrations so that, from May onwards, it can also be found in the northern regions of Europe. With strong tailwinds, the hawk moth can reach a flying speed of up to 100 km/h. Without any winds, the hawk moth can fly briefly at a maximum speed of 60 km/h, but even then it qualifies as an extraordinarily fast flyer.

Kritsky G. **Darwin's Madagascan hawk moth prediction**
American Entomologist 37, 206-210 (2001)
Wasserthal L.T. **The Pollination of the Malagasy Star Orchids** *Angraecum sesquipedale*, *A. sororium* and *A. compactum* **and the Evolution of the Extremely Long Spurs by Pollinator Shift**
Botanica Acta 110: 343–359 (1997)
Wasserthal L.T. ***Angraecum*-Orchideen und langrüsslige Schwärmer, Bestäubung und Evolution**
Die Orchidee 66(3), 175-181 (2015)

"Cute predators"

Within the class of insects, dragonflies have developed a unique form of mating. First, the male grabs the female behind her head, clasping her neck with his anal appendage. If he succeeds, the female cannot escape. The female will bend her abdomen towards the basis of the male's abdomen in order to meet his secondary genitalia. The pair of dragonflies thus forms the famous heart-shaped mating wheel, like the one pictured here, which is formed by two specimens of the white-legged damselfly *Platycnemis pennipes*. Prior to this, the male must charge its secondary genitalia with sperm in order to guarantee a successful transfer of sperm to the female's genital opening.

Pfau H.K. **Functional Morphology and Evolution of the Male Secondary Copulatory Apparatus of the Anisoptera (Insecta: Odonata)**
Zoologica 156, 103 Seiten, Schweizerbarth Verl, Stuttgart (2011)

Pure action

Highly active during the midday heat

Observing insects in their hustle and bustle during the midday heat can be quite rewarding for an animal photographer. However, you have to be quick on your feet and ready to shoot: Many an exciting scene takes place "in time lapse."

Left: A wasp of the genus *Sceliphron* sits on a flower. All of a sudden a second wasp seems to appear out of nowhere and approaches the first wasp. The two wasps appear to become engaged in a battle. However, in most cases, they are only curious to find out if the other is a female or a rival.

Within one sixth of a second

Two flesh flies of the genus *Sar-cophaga* (image to the right-hand side, upper half) are obviously ey-eing each other. The camera shoots 12 images per second. In the next image, the two flies are already in-volved "in a scuffle" (bottom left). A further twelfth of a second later (bot-tom right), it is barely possible to ma-ke out which and where the two si-des are.

Mating ritual

The two flies pictured in the image series at the bottom perform a nup-tial dance, which lasts some see-mingly never-ending 10 seconds: First, they position themselves paral-lel to each other, then they look each other deep into the eyes, and finally engage in copulation.

On a knife edge

Like the forced-guiding mechanism of a drawing device

Preparation phase for plunge diving: The bird first folds its wings together so that they form the shape of a capital M. This is enabled by a kind of parallel-guiding mechanism in its arm skeleton. The upper arm bone is flexibly linked to the merged hand bones of the wing via the ulna and radius bones in the lower arm. When the wings are drawn to the body, the distal portion of the wing lays itself automatically across the proximal portion.

Blue-footed booby (*Sula nebouxii*)

The wing tips touch the water

This is actually quite unusual, but the booby does not mind, because its plumage is well oiled.

The moment of immersion into the water is critical
The booby continues drawing its wings in until it is shaped like a bullet, similar to a peregrine falcon during a nosedive (see page 72f.). The tail feathers are spread to their maximum, because they must ensure that the bird dives into the water along its longitudinal axis. Just the slightest tilt would cause the booby to break its cervical spine. Shortly before the bird plunges into the water, the cervical vertebrae are reinforced through tensioning of the muscles.

Steep upstroke during a short flight on the spot
The images clearly show how the covert feathers curl in several places. The flow separation point travels from the back to the front, but the spread feathers avoid an immediate collapse of airflow and thus a sudden drop of the flying object. In the images, it can also be seen how the feet with their spread webbing are used as navigation surfaces during flight.

Exciting if you look closely

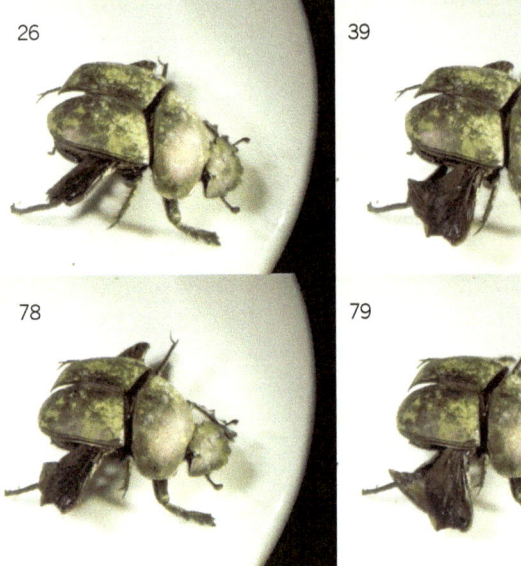

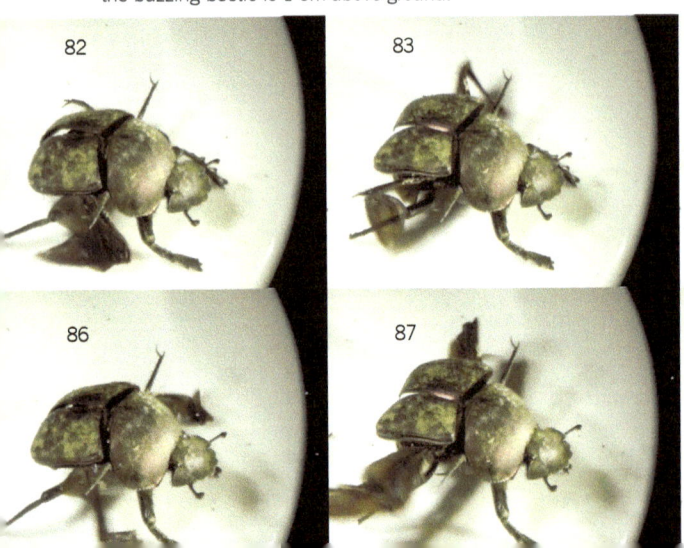

The folded wings are usually fully extended first

and only then do they start beating. This may be observed, for instance, in the stag beetle *Lucanus cervus* pictured on the right-hand page. In the case of the scarab, however, it is only after the first wing-beats that its wings are fully extended. This is quite unusual. The centrifugal forces generated in these moments may cause the outer third of the wings to unfold quickly.

The stag beetle's flight preparation takes longer

After the opening of its elytra and the slightly delayed unfolding of its wings, the stag beetle seems to hang on, "doing nothing" for a while as well (the stag beetle may remain in this nearly motionless position for up to a whole second longer than the scarab). Only the front legs seem to dangle helplessly in the air as the beetle brings itself into as upright a position as possible. Then the wings begin to beat unexpectedly fast. This often catapults the heavy beetle almost vertically into the air (see page 64f.).

Take-off of a scarab *Scarabaeus* at 250 frames per second

In the upper image series, images 0, 13, 26, and 39 are shown (that is, the images were taken at an interval of 1/20 second). Not much happens during this "relatively long" time span: The wings are folded out halfway. Then there is even a "break" of one sixth of a second. Things finally start happening from image 76 onwards (that is, 0.3 seconds after the first action, second image series from the top): The wings, which are not fully folded out, are vehemently raised and lowered, making 125 (!) beats per second.

The lower two image series show images 82-89 (covering a time span of merely 1/30 second). During phase 82-85, the wings are fully extended as they continue beating at a frequency of 125 HZ. The beetle begins to rise into the air from images 86-89. In images 98 and 100 on the right-hand page (0.4 seconds after the first wing movement), the buzzing beetle is 1 cm above ground.

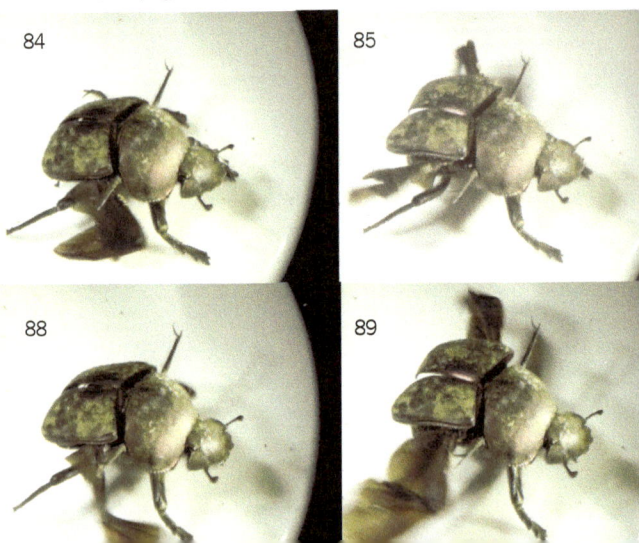

Extreme acceleration

The wing tips of both the scarab and the stag beetle can reach an enormous absolute velocity (up to 15 m/s). Even more astonishing is the acceleration, as well as delays, produced during this process. Let us look at images 98 and 100 from the high-speed photography series. An interval of 1/125 second lies between the two images. During this interval, the beetle beats its wing back and to the front again. If the beetle accelerates from the frontal zero position to the central downstroke position (10 m/s) within one quarter of that time (that is, 1/500 s), the wing tips experience an acceleration of 500 g.

Turning the corner

The angle of inclination does not depend on the mass

Every animal – and also every technical object such as a bicycle or an airplane – must "lean into the curve." The image on this page (multiple photographs) shows a leaf-nosed bat Carollia perspicillata making a sharp turn. The wings' angle of inclination α depends solely on the velocity v and the curve radius r, but not on the mass: $\tan \alpha = \frac{v^2}{gr}$. A swallow must lean into the curve to the same extent as a commercial plane. Conversely, if the curve radius and the angle of inclination are given, it is possible to calculate the velocity. For instance, $\alpha = 45°$ and $r = 0.4$ m equals a velocity of $v = 2$ m/s (7,2 km/h).

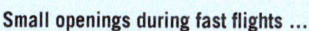

Small openings during fast flights ...

Unlike common house martins, barn swallows *Hirundo rustica* nest inside of buildings. To enter these buildings, these masters of flight acrobatics do not require more than an opening the size of a beer-mat, through which they slip quickly – with their wings momentarily tugged in – during flight. They catch their prey, flying insects, in free flight. While foraging, they often fly quite low, depending on where the insects they are looking for are found, such as over the surface of lakes. The flight silhouette of adult specimens is unmistakable due to their long, deeply forked tails.

Barn swallows are typical migratory birds

They commute between their nesting areas in our regions and their winter habitat between Equatorial Africa and South Africa. During their migration, they follow a set route. The barn swallows from England fly over West Africa, the ones from Eastern Europe over East Africa, and Norwegian barn swallows fly over Central Africa to the tip of South Arica. There are also several cross-connections between these routes. Swallows travel some 5000 km at least twice a year, crossing 3000 km with each trajectory over the Sahara and the Mediterranean.

It may be cooler up in the air, but there are often headwinds

It was previously assumed that, like the common swift, swallows crossed the Mediterranean and the Sahara in one go, flying non-stop at great heights for 40 hours. Today it is believed that they might have a migration strategy similar to that of the typical warbler, which migrates in stages, flying at night at an altitude of only a few hundred metres and spending the days in an oasis in order to escape the heat. It may be cooler high up in the air, but there are often headwinds, so migrating birds must always assess their current situation.

Being spoilt for choice

A subspecies of the American yellow warbler *Dendroica pete-chia* ssp. *aureola* from the Galapagos Islands has found a spot with many flies flying around. It skilfully catches one of the flies with its sticky tongue, but then spots another fly nearby. You can see how the bird is facing a dilemma, as flies are very fast …

The African spoonbill *Platalea alba* has caught a giant water bug *Belostoma*. The insect tries to defend itself desperately and slips through the bird's grasp several times. Eventually, the spoonbill manages to swallow its "big meal." The giant water bug reaches a length of over 10 cm and probably weighs 20 g.

A complete wingbeat

May bug (*Melolontha melolontha*)

The 50-60 Hz-range

Left-hand page: a May bug in flight, filmed at 500 frames per second. Ten consecutive images show a complete wingbeat (images 1-5 show "the lower half," images 6-10 "the upper half"). The large bug, which reaches a length of about 25-30 mm, thus beats at a frequency of 500/9 Hz, that is, about 55 Hz. Hummingbirds and moths also fly at this range of frequency (see p. 16). Within that short period of time, the bug manages to move 6-7 mm, which equals a velocity of about 3 m/s. Its elytra flap concurrently with the wings, like that of its relatives (see p. 194), and its body position is almost vertical (see p. 67). Its antennae, which vibrate like the elytra, are fanned out widely: They are used for orientation through scent recognition. The male pictured on this page sports seven antenna "leaves" with some 50.000 olfactory nerves, which help to detect females.

The 2 Hz-range

On the right-hand page, there are four consecutive images taken at an interval of one twelfth of a second. The images capture a scene in which two rivalling tufted ducks chase each other. "In sum," you can see more than one complete wingbeat. So the 45-cm long duck beats its wings about twice per second.

Size and frequency

A comparison of wingbeat frequency yields a ratio of almost 30:1 in favour of the bug; a comparison of length yields a ratio of 1:15. In general, the following principle applies: Larger flying animals have a lower wingbeat frequency than smaller ones, ranging from 1 to 1000 (see p. 52).

The Reynolds numbers (see p. 56) tend to be lower in small animals, which does not allow for immediate comparisons of their fluid mechanics.

Tufted ducks (*Aythya fuligula*)

Index

Scarce swallowtail (*Iphiclides podalirius*)

Picture credits: Drawings/paintings pp.xi, 7, 10, 11, 12, 13, 14, 54, 72, 79, 81, 83, 125, 166, 170, 171, 172, 173, 175: Markus Roskar photos p.108,109, 215 lower image: Hannes F. Paulus, all residual photos, computer drawings: Georg Glaeser